D1736774

Hazardous Material (HAZMAT) Life Cycle Management

Corporate, Community, and Organizational Planning and Preparedness

Hazardous Material (HAZMAT) Life Cycle Management

Corporate, Community, and Organizational Planning and Preparedness

ROBERT JAFFIN

CRC Press
Taylor & Francis Group
Boca Raton London New York

CRC Press is an imprint of the
Taylor & Francis Group, an **informa** business

CRC Press
Taylor & Francis Group
6000 Broken Sound Parkway NW, Suite 300
Boca Raton, FL 33487-2742

Printed in the United States of America on acid-free paper
Version Date: 20120823

International Standard Book Number: 978-1-4398-7387-8 (Hardback)

Library of Congress Cataloging-in-Publication Data

Jaffin, Robert.
Hazardous material (HAZMAT) life cycle management : corporate, community and organizational planning and preparedness / Robert Jaffin.
p. cm.
Includes bibliographical references and index.
ISBN 978-1-4398-7387-8 (hbk. : alk. paper)
1. Hazardous substances--Management. 2. Hazardous wastes--Management. 3. Materials management--Environmental aspects. 4. Business logistics--Environmental aspects. I. Title.

TD1030.J34 2013
363.17--dc23 2012031515

Visit the Taylor & Francis Web site at
http://www.taylorandfrancis.com

and the CRC Press Web site at
http://www.crcpress.com

This book is dedicated to the silent majority: the hundreds of thousands of people who remain unseen and unknown but who make our system work. That includes everyone: the able-bodied seamen and oilers; the commercial motor vehicle enforcement personnel who crawl underneath huge 18 wheelers; and the more visible but seldom recognized people, such as industrial hygienists. It includes all those people on the front lines in the supply chain system who keep everything going and moving. For many years in the Department of Defense (DOD), we would talk about the "little old ladies in tennis shoes" when we talked about the vast defense supply system. That phrase referred to all the unknown and unrecognized civilian employees who keep things moving. It is their wealth of experience and their shared lessons learned that allow senior management and senior designers to improve products and processes. The vast silent majority are all those behind-the-scenes people who allow global enterprise to continue in what usually appears to be a totally seamless manner. Without the lessons learned and shared by all those people, supervisors and managers would never be able to correct problems.

In the military it has long been recognized that the "tip of the spear," the service member on the very front lines in case of conflict, is supported by a complex and convoluted network of individuals typically with 11 or more identifiable positions supporting each front-line service member. In the commercial sector, the maintenance personnel, janitorial staff, material handling equipment operators, warehouse workers, shippers, and receivers are often forgotten, but without the efforts of every one of those people, nothing would happen. This book represents lessons learned from all of those front-line workers, as interpreted, filtered, and reported by their chain of command.

Contents

SECTION II DEVELOPING PROGRAMS FOR BROAD SECTORS

Preface

This book was written to provide a solid foundation for the organizational management of hazardous materials and wastes. As a result, it focuses on managerial and strategic issues and attempts to highlight the impact of strategic decisions on the actual operating environment and on society as a whole. An introduction to the concept of hazardous materials, including classification, unique terms, critical definitions and concepts, and the regulatory framework (Chapters 1–3) comprise Section I. The commercial sector, the government sector, and nongovernmental organizations (NGOs) are treated separately (Chapters 4–6), and how to put it all together (Chapter 7) makes up Section II. This text is slightly different from many others because of the extremely large appendices. Due to the sheer size of the appendices, they have been placed on the accompanying disk included with the hard copy editions of this book. The appendices offer rapid access to excerpted portions of the most critical parts of key regulations and standards.

This book is the result of three intersecting drivers. For my first and only doctoral research course I selected hazardous material management as the starting point for my eventual dissertation topic. While doing research and extensive literature searches and reviews throughout the course, it became clear that there has been very little written on this topic from a managerial and proactive perspective, and what little has been published dates back 20 years or more. In the last six years, I developed and delivered four graduate and undergraduate hazardous materials management courses, and there was not an appropriate core or basic text for any of those courses. While attending the FEMA higher education conferences and networking with large numbers of academicians and professionals, it became apparent that there is a dearth of material available related to hazardous materials management, rather than managing the hazards of hazardous material, and that there was genuine interest in seeing someone (me) finally sit down and write an updated text. It is really the support that I got at the FEMA higher education program conferences that led me to this undertaking. The book has been in the "idea" folder since 2002. Two and a half years ago, as a result of the encouragement I received at the FEMA higher education conference, it came out of the folder and became a working document.

It is designed as an integrative baseline document that creates common ground and uses common language to allow everyone from CEOs to janitors to better understand hazardous materials at the strategic level, as well as the inherent dangers of hazardous materials—even when they are sitting in a storeroom that has to be mopped once a week. That means the audience is every HAZMAT employee and every HAZMAT employer. Within the framework of higher education the material here has value for career tracks in business, business continuity management, civil engineering, defense management, emergency management, environmental science, homeland security, industrial hygiene, public safety, public health and administration, security management, and transportation logistics supply chain management.

I would be remiss if I did not thank a number of individuals who have helped and encouraged me throughout the process: My wife Laine has always been there for me and rooted me on, and Mark Listewnik was ready to take on this project and keep me on track. I also want to thank Wayne Blanchard, Lucien Canton, Patricia Eisenstein, Oliver Hedgepeth, Claire Rubin, and Christine Springer; they all provided support as the book developed.

About the Author

Robert Jaffin has 30 years of combined active and reserve military service, 11 years of experience in analytical instrumentation sales and service, and just over 30 years of experience in transportation and logistics. He has called on chemistry laboratories in hospitals, university research facilities, state police laboratories, and state environmental laboratories. He has also called on federal research facilities and both production QA labs and research laboratories in segments that include steel making, power generation facilities, animal food producers, and fertilizer producers. Jaffin has lectured in 16 countries and taught short professional or technical courses in 25 countries. He holds a bachelor's degree in business and an executive MBA.

Jaffin has taught graduate and undergraduate business courses, hazardous materials management courses, and transportation logistics courses, as well as emergency management and homeland security courses at the undergraduate level. He has lectured at the United States Merchant Marine Academy, at the Massachusetts Maritime Academy, and in the healthcare emergency management program at Boston University School of Medicine. Jaffin has also staffed and took through accreditation six undergraduate and six graduate degree programs: Emergency and Disaster Management, Environmental Science, Fire Science, Homeland Security, Public Health, and Transportation Logistics Management. He was a HAZMAT transportation certifications instructor for the Navy from 1991 to 2000, where he developed four new courses and delivered on-site training to all DOD components in the United States and overseas.

Jaffin has written regularly for professional journals and served as a manuscript reviewer for publishers such as McGraw-Hill and John Wiley & Sons. He sits on the Training and Education and the Food and Agriculture committees of the International Association of Emergency Managers (IAEM). He has been an active participant at the FEMA Higher Education Conference since 2003; in 2008, he organized and put on their first-ever full day writers' workshop, which is now part of the annual conference. He sits on multiple Transportation Research Board panels and committees and is trained as an Homeland Security Comprehensive Assessment Model (HLS-CAM) instructor. Jaffin serves as one of many commissioners on the Stratford Regional Planning commission in New Hampshire. From 2002 until 2004 he provided on-site HAZMAT awareness training at multiple field and classroom locations to over 600 reserve POL drivers bound for Iraq and Afghanistan.

In the past, Jaffin has served in such diverse roles as committee chair for the Exercise Planning Committee of the Jefferson County West Virginia LEPC, HAZMAT awareness trainer for the county CERT program and the sheriff's office, member of the county counter-terrorism committee, and homeland security instructor for the National Sheriff's Association. He has served as a panelist at the International Air Transport Association (IATA) International Dangerous Goods

Conference in Atlanta, Georgia; as a panelist at the National Defense Transportation Association (NDTA) European Symposium in Stuttgart, Germany; and as president of the National Defense Transportation Association, Atlanta chapter.

INTRODUCTION AND DEFINING THE KEY TERMS USED DOMESTICALLY AND INTERNATIONALLY

I

Chapter 1

Introduction, Background, Definitions, and Conversions

February 22, 2012

Hechi, China
River poisoned: A massive spill of toxic cadmium has been coursing down the Longjiang River in southeastern China for at least two weeks, threatening the drinking water supply of millions of people. Cadmium, mined in the region for use in batteries, can cause cancer if ingested. When it was first detected in the river, in mid-January, officials suspected a mining company that had repeatedly been cited for waste-disposal violations. But since then seven officials from various chemical companies have been detained for questioning, and the government ordered at least six other mines and factories to suspend operations temporarily. The city of Hechi and surrounding areas are known to be contaminated with cadmium, lead, arsenic, and other heavy metals.

1.1 Overview and Background

This work attempts to fill the gap related to business and hazardous materials as a unique but extremely diverse group of materials that must be treated within the framework of supply chain management. An overarching framework that truly recognizes the precursors: Logistics engineering and the current supply chain management concept. Essentially, this starts with the original concept of logistics engineering that was so well defined and addressed by Blanchard (Blanchard, 1974). It builds upon many of the concepts adopted by the US and British military forces under what was known as Military Standard 1388, or MILSTD1388, a standard that addresses life cycle management. It builds upon the increasingly regulated world but also recognizes new and growing disciplines such as the burgeoning field of reverse logistics and growing interest in "green" models and sustainability.

This work focuses not on the hazards of materials, although the intrinsic issues related to those facets is not ignored, but rather on the elimination or the minimization of their generation and use and the application of appropriate risk management techniques to minimize corporate liability and exposure, as well as to minimize impacts on the as-built environment. Another key concept is the intrinsic understanding of why we must be concerned whether or not there are currently adequate regulations. Perhaps the easiest example to see is the concept of throwing a lighted match into a can of gasoline or smoking while refueling. You do not need the government or a set of regulations to understand why one should not do that. This text addresses the business concept of hazardous material and the understanding of what hazardous material really means at that basic level. This is not about the regulations but a much deeper understanding of why we need to take these actions. Since actions today must be guided by, or grounded properly in compliance with various state actor regulations, we will depend very heavily on those regulations to help us identify materials that either present hazards as we think of them or are regulated because of theoretical—although perhaps highly unlikely situations—where they would become hazards to the global environment.

While this text does not directly address reverse logistics, recycling, and recovery, they all have become important parts of the supply chain and are addressed in supply chain management, logistics management, transportation management, and reverse logistics management. In its broadest sense, reverse logistics can be seen as a revised and updated life cycle management or logistics engineering approach. There is a growing understanding of the negative impact that industrial wastes—solid, liquid, and gaseous—have on the environment. This text will explore this area as part of the larger environmental health and safety construct (with some discussion of the global ecosystem), a discussion that goes far beyond what the transportation regulations address, although we use those transportation regulations as the framework for developing a large portion of the material here.

Informal estimates place *the number of hazardous materials movements within the United States as upward of 1.2 million movements per day*. That does include some materials that get moved more than once in a single day, typically by multiple modes. That number is a statistical indicator of how much regulated material is actually consumed on a daily basis. Unfortunately, that does not cover many materials that are of concern and are essentially discharged into the ground into aquifers or into the atmosphere. Many hazardous materials are not recognized at the retail level, by stores, employees, or customers, because, although they are regulated, they are not considered hazardous materials for shipment and retail sale.

Some of those same materials become regulated as waste within a commercial setting but not necessarily in the retail or private/individual user universe. That is changing, and some states already have, or are in the midst of, enacting regulations that do impact private users and homeowners. In the past, there has been a dichotomy between nonretail and transporter responsibility in regard to regulated materials and wastes, and retail outlet and private citizen responsibility for these same materials' wastes. Some factors that combined to drive this include: an increase in consumption of material goods that are more technologically complex and contain more hazardous materials; a switch from products that represent no hazard for disposal to newer products that must be treated as hazardous wastes (specifically the move to replace incandescent light bulbs with fluorescent lights and the newer compact fluorescent lights or CFLs); and, a growing body of knowledge that demonstrates the need to more effectively manage waste streams via statutory requirements. Today, in both retail trade and for private consumption, primarily, but not exclusively, in regard to waste disposal of a very large number of consumable products that were not subject to regulation in the past (as well as new products being phased into common use by government fiat) have become—and are becoming—subject to regulation. Individual workers and individual consumers can no longer avoid responsibility for proper handling and disposal of wastes, hazardous or otherwise. If one includes retail products that are regulated but not classified as hazardous—materials that are not hazardous in their packaging or when used properly but become hazardous if they leak, break, or are improperly used or disposed of—the numbers would triple to quintuple. To demonstrate just how many end-use materials generate, or must be treated as, *hazardous waste*, here are a few examples: all fluorescent light bulbs, all batteries, many household cleaning products such as ammonia and bleach, and most insecticides and pesticides as well as many fertilizers. To the consumer, these are products one expects to see in most retail outlets but to the manufacturers and commercial sellers and resellers, they are often regulated items.

A comprehensive and well-documented set of standards is currently applied to all business and commercial applications relating to both hazardous materials and hazardous wastes. Currently, there is a lack of standards applied to private use and disposal, and even where states have regulations on the books, they fail to fund outreach education programs or collection programs, so the public at large is often unaware of the requirement to properly dispose of material and how or where to dispose of their regulated wastes. That huge disconnect between commercial application and personal consumption, when coupled with the failure of many in business to recognize the difference between the concept of regulated materials and materials clearly identified as hazardous, tends to blur and confuse the issues for the majority of the population.

The key here is to grasp the concept of hazardous materials so that one may apply it to any product and any application in the commercial, governmental, nonprofit, and private sectors. A first step is to make sure that one understands the language in the various hazardous materials or dangerous goods regulations. Americans also have a language disadvantage compared to the rest of the world as the United States is the only country that considers the term *hazardous material* the primary identifier, but in every other country, including Canada, Western Europe, and Asia, the correct term—derived from the United Nations Economic and Social Council's Committee of Experts on the Transport of Dangerous Goods—is *Dangerous Goods.*

A Google search on the phrase *hazardous materials management* produces approximately 967,000 hits. I have scanned the first 10 pages of results and identified three that address the topic of hazardous materials management as part of business and logistics management and of those, two refer to a particular standard. Every other link addresses technical issues and regulatory issues, including but not limited to compliance, related to the dangers of hazardous material and the physical sciences as distinct from business logistics and operations management or operational research.

As we will learn, very few hazardous materials are generated or produced in one location and then consumed at that same location. Therefore, transportation becomes the critical element or enabler regarding hazardous material. In addition, until very recently the only international body of work that addresses hazardous materials in an organized and recognized fashion is the transportation regulations developed out of the United Nations recommendations commonly referred to as the "Orange Book."

Pollution's largest identified contributor in the United States is what is referred to as *nonpoint source* pollution. That in turn is made up of two components: chemical runoffs and transportation prime mover emissions.

1.1.1 WHAT DO WE MEAN BY HAZMAT AND HAZMAT MANAGEMENT?

1.1.1.1 Key Definitions as Applied in the Text

1.1.1.1.1 Basic or Bedrock Definition of Hazardous Material = FED-STD-313D With Change 1

There is no simple or concise definition for the term *hazardous material*. The federal government uses that term in many different regulations, and it is defined differently in all of them. If one wants to get a reasonable definition of the term, one is forced to look not at the federal regulations but at the federal manuals for doing business. That serves to underscore the general misunderstanding of federal regulations and the generally misunderstood domain of hazardous materials. The underlying issues are societal and environmental, but common sense and proper business and management theory apply much more than the continued focus on the operational issues related to addressing the dangers of hazardous material. That is the underlying rationale for the creation of this text. In order to define, understand, and deal with hazardous materials, we need to first understand basic business processes and management concepts and then we need to understand what we really mean by hazardous material as distinct from regulatory requirements or how to meet compliance goals and objectives. The use of hazardous materials within society is based in the notion of trade-offs in terms of safety, cost-efficiency, and similar objectives. The bulk of those are business organization objectives not environmental or regulatory objectives.

If one wishes to do business with the government, one must comply with the Federal Acquisition Regulation, the FAR. There is a subset or variation of these regulations known as the Defense Federal Acquisition Regulations, or DFAR. Referring to either of these regulations will eventually lead you to a "derived" definition for hazardous materials as it applies to both the business of government and the enforcement of government regulations. Both of these documents state that the definition of hazardous materials for the purpose of doing business with, or for the government, and, for all activities within the federal government, is defined in Federal Standard 313. The most current version of the standard is FED-STD-313D dated April 3, 1996, with one change dated March 21, 2000. The agency responsible for the standard is the General Services Administration (GSA). Here again, notice that the basis for control of the definition falls within the business functions as distinct from the governance or regulatory functions of government.

The process one needs to go through to identify where the federal definition of hazardous material lies, to begin to fully understand the ownership of that standard and to its lack of currency, are all indicators of why this subject is so poorly understood and so rarely treated properly as a business consideration rather than a regulatory requirement. The definition falls within the acquisition regulations, which are operating documents, and the defining document is a federal standard that falls within the GSA, an organization that is focused on the purchase of goods and services, not scientific and technical definitions. Notice that the underlying root document has nothing to do with the hazards of hazardous material, emergency response, personal protective equipment, or the myriad of issues most people usually associate with the term hazardous material. They talk about business processes and business definitions.

1.1.1.1.2 Hazardous Substance = EPA 40 CFR

We will use the abbreviated but highly effective definition found in FED-STD-313: *hazardous substance means any substance designated pursuant to 40 CFR, Part 302, Table 302.4.* Recognize that this standard has not been updated since March of 2000, nor has there been any notice of replacement or withdrawal, so there may be additional locations and additional material that now fits this definition. You will find a fuller discussion of the acts, or laws, that drive these definitions and drive the creation of the 40 CFR. For the purpose of the text, we will use the tables that were created for 49 CFR § 172.101 appendices; you will find them in Appendix A7.4 of this text.

1.1.1.1.3 Hazardous Waste = EPA 40 CFR

We will use the abbreviated but highly effective definition found in FED-STD-313: Hazardous waste shall have the meaning provided in 40 CFR Part 261. Recognize that this is a standard has not been updated in 12 years so there may be additional locations and additional material that now fits this definition. You will find a fuller discussion of the acts, or laws, that drive these definitions and drive the creation of the 40 CFR. For the purpose of the text, we will use the tables that were created for 49 CFR § 172.101 appendices; you will find them in Appendix A7 of this text.

The physical characteristics as well as chemical characteristics of the material may impact whether it is classified as well must be treated as one of all of the above. And, in fact, that can change exactly how the material is classified. A material may meet all of the above definitions or only one of the definitions. These are not usually inclusive or mutually exclusive other than the fact that every hazardous waste by definition falls within the larger category of hazardous substances. Some classic examples of materials that get treated differently although they might appear to be the same include such things as wet hay and dry hay; a solid piece of beryllium and beryllium dust, and solid or embedded asbestos versus "frangible" asbestos. There are many other examples of these, which should give you some insight into the complexities of both classifying and handling materials. A sheet of lead is not considered hazardous materials, but lead dust or lead powder most certainly is if it becomes small enough to be "aerosolized" or inhaled.

1.1.1.2 Hazardous Material Management

There are many differing definitions but almost without exception the concept of hazardous materials management—as discussed and defined in the majority of texts currently available, including material used over the last 20+ years—focuses on the management of the hazards of hazardous material not the overall or life cycle management of hazardous material. For the purposes of this text, the term *hazardous materials management* refers to the business and life cycle management of all business, industrial, production, and, janitorial and maintenance processes that generate products or streams regulated as hazardous. This includes all processes from concept development to product development to final product disposal. It specifically addresses discharges into the atmosphere, discharges of solids or liquids into the landmass including bodies of water, and includes disposal of material into solid landfills or other destructive processes that do not neutralize or remove the hazards related to such materials.

1.1.1.3 Other Key Terms and Concepts Necessary to Understand the Hazardous Material Universe

There are other key terms and concepts, and their definitions are far less important from a practical perspective than is an understanding of why they exist and what they represent. Under the various international and national dangerous goods and hazardous materials transportation regulations, there are a number of systems used to categorize groups of materials that scientifically or technically represent some inherent level of risk but that do not require identification and compliance with the bulk of the global hazardous materials regulations.

1.1.1.3.1 Regulated Material

This term has multiple definitions, but all of them are close and some are merely unique to a particular regulation without drastically altering the concept. If the material requires an MSDS, then it is by definition regulated. If the material requires special handling in the commercial sector or during transportation, it may become regulated. At the same time, many such materials are available in every sort of retail outlet. For some of those materials, each retail location is required to maintain MSDS sheets that must be made available to the buying public if requested, and meet other environmental health and safety standards. Some materials require an even lesser standard once one gets to the retail side. Those items appear on store shelves in supermarkets and convenience stores, barber shops and beauty salons, the vehicles of almost every building contractor, and other outlets where even the reseller is not required to maintain documentation. In many cases, those resellers are subject to multiple regulations and are not even aware that they are subject to the regulations. An oversimplified example would be pressurized deicer or WD-40: if you carry them in your car and use them during your commute as a private citizen, you can consider them unregulated for transportation. On the other hand, if you are a school bus driver, truck driver, or a contractor, those items fall into a special category known as materials of trade and are subject to some very specific, but basically minimal regulatory requirement. The concept and application of "materials of trade" applies to every law enforcement vehicle, every fire service vehicle, every ambulance, and every public works and highway department vehicle as well as to all privately operated tow trucks. As late as 2000, the US military, large segments of the uniformed public safety community, and civil government in general fail to recognize or accept that they are subject to the regulations. Today, most local entities still do not understand that they as organizations and their members as individuals are subject to the regulations. In essence, the problem is more properly laid at the feet of the federal government but, conceptually, federal government and all lower levels of government need to rethink their approach in these areas and move from a regulate to an "educate and regulate" mode. Some of those materials would also fall under the special category of company-owned material (COMAT) which while regulated is subject to sometime conflicting or confusing application of the regulations.

Another distinction between regulated materials and hazardous materials, as a generic term, is that many of those materials are regulated for disposal, but the government has never educated the public nor made appropriate arrangements for the reverse logistics required to support the recovery recycling effort. One of the best examples of hazardous materials that are regulated yet at some point get "lost" in terms of life cycle management are batteries. Some batteries represent an extremely high risk and are regulated down to the retail outlet. Other batteries are considered to represent a much lower level of risk and are regulated in essence only through the wholesale chain. Other batteries represent such a low risk that they are not recognized as hazardous materials after their production, although MSDS sheets are available. All however, from a practical standpoint, should not be disposed of but rather recycled. Lithium-ion batteries have caused the loss of at least two large commercial aircraft, led to serious fires in homes, and have led to numerous laptop PC fires. On the other hand, many items that are intrinsically hazardous materials were ignored or not subject to regulation because the full impact and cost of these materials as waste streams was not understood or appreciated. One example would be all with different forms of what in the past would have been referred to as "dry cells" batteries; they are not subject to any form of effective regulation or recognition through the majority of their life cycle, they do represent a real, if low-grade, risk as they are based on an acid reaction and therefore they are inherently corrosive and therefore hazardous to the environment when they decompose as part of waste site effluent and leachate. The more pervasive and widely used items include ammonia bleach, all aerosols, and, many HBA items including hydrogen peroxide. The bulk of these materials have been reclassified within the United States to a very broad category currently identified as other regulated material–domestic, or ORM-D. The increase in severe weather incidents that lead to flooding and overflows of treatment plants/facilities and landfills is an issue that deserves a great deal of additional research and study.

Another example would be fluorescent lights. With the US government's push to eliminate incandescent bulbs, this has already become an issue that is causing many problems. From a business or commercial standpoint, the dangers fluorescent bulbs pose have been recognized for a long time. The requirement to recycle rather than to dispose has led to a small cottage industry merely for designing developing and delivering packaging to allow these used or bad units to be returned into the system for proper disposal. Newer compact fluorescent lights are designed specifically to replace a large majority of incandescent light bulbs; they are just as hazardous as traditional fluorescent light bulbs. The EPA has published disposal directions for the public and has worked with manufacturers and retail chains to set up programs and facilities to accept used or burned-out CFLs as well as fluorescent light bulbs. Many hardware store chains also accept batteries for recycling as well. This is perhaps the most striking example of government regulation gone bad. No one, including private citizens, has the right to throw out any form of fluorescent light bulb. Yet it is safe to say that as of 2010, 90% of the population in the United States is not aware of the fact that these bulbs must be recycled through an authorized facility and that they represent a very real (if minimal) inhalation and contamination risk when broken in a household. The nature of the EPA guidance for disposal of these items is the first issue here. In essence, the EPA says it is your responsibility to know that you cannot throw the bulbs out, that is your responsibility to check with both your local municipality and the state to find out how to dispose of such materials, and, if you cannot get that information how to double bag such items in plastic bags so that they may go to the landfill. If you place fragile glass light bulbs of any sort in double wrap plastic bags and then throw them in the trash, those bulbs are likely to break and the sharp edges are likely to puncture the wrapping. That yields additional plastic trash bags that will not decompose and are no longer fulfilling a useful purpose, and, mercury or mercury vapors that are building up in our solid waste facilities both as gases and as absorbed contaminants to liquids. There are now a growing number of commercial efforts to recover methane from solid waste disposal facilities. If we continue to allow disposal of mercury containing light bulbs in such facilities, we are contaminating a potential source of fuel that will force us to deploy expensive "scrubbing" processes/hardware and software to remove any traces of mercury, thus negating many of the inherent economic advantages of such recovery. There are similar arguments and considerations for many other consumer wastes we do not presently educate on or adequately regulate at the end of their life cycles. At the current time, almost none of these issues are being addressed. In essence, this guidance makes it easy for private citizens to continue disposing of fluorescent lights improperly and offers those citizens a technique that compounds the issue rather than resolves the issue. The second issue with the federal government's approach and the EPA guidance is that it has been done in the traditional "shove it down your throat" manner as a government edict rather than a government-sponsored education program that starts in the school systems and includes considerably more focus on outreach rather than enforcement.

Many lubricating oils are also subject to regulation. The dangers of waste oil were recognized a long time ago, and these materials are now regulated through their entire life cycle. Even individual consumers are held responsible for proper recycling/disposal of such materials. Tires are yet another product that is being managed on a life cycle basis with proper disposal of tires being a major industry. There are significant dangers with the older technique of aggregating large numbers of tires together for recycling. One need only consider the tire fire under I-95 in Philadelphia in 1996 to learn about unintended consequences and low-probability high-impact incidents as well as generation of toxic or hazardous clouds or fumes.

In the United States, many of the leading hardware and discount chains have active recycling programs; in many cases they offer to recycle products bought at the store. Think about how consumers would be able to identify batteries and light bulbs, audit particular stores, or how stores can enforce such restrictions. On the other hand, unless the government is providing direct financial support or offsetting credits of some sort, how or why should such outlets put themselves at such high risk at such high cost? This question too is considered whenever it arises in this book.

There are many products that are in essence "regulated" while not being classified as "hazardous materials." Standard "dry cells" are based on a reaction using either an acid or an alkali. For transportation purposes and for retail distribution, including sales, they are not considered hazardous materials; nonetheless, their movement does fall under the category of "regulated" materials. Batteries present the best example of hazardous materials. Every battery currently in use represents a hazard of some sort, but many are not classified as hazardous materials and even fewer are recognized as hazardous waste. The lack of standards and awareness in the business community, as well as in the individual consumer community, continues to add to documented environmental concerns. Another example is all of the older fluorescent tube lights in any configuration and the newer CFLs. There are disposal requirements for private individuals as well as all commercial entities, but both the hazard of these devices and the proper disposal techniques have been "hidden" from the public primarily because of political concerns. Items as diverse as "whiteout" or typographical correction liquids and dry erase markers have been regulated at one point or another. Chemicals that once were seen as saviors in agriculture and other parts of the economy are now banned as carcinogenic or for other hazardous characteristics.

Misinformation, disinformation, lack of information, and a pervasive failure of the public to understand what are hazardous materials and the need for proper disposal techniques has resulted from uneven enforcement actions and from the conflicts between definitions from organizations such as OSHA, EPA, and DOT.

There are many different groups of products and materials that are technically or in fact are hazardous materials and are subject to the regulations, but most of these materials are exempt from large portions of the regulations. Typically, producers, shippers, and receivers of bulk wholesale quantities of these materials must be aware that these materials represent a theoretical hazard, but they also must recognize that the considerable portion of the regulatory requirements, especially those related directly to "hazard communication" as defined in 49 CFR Part 172 subparts C, D, E, and F, are not applicable in the retail marketplace or during transportation. Some examples include most alcoholic beverages; medical samples, including blood and urine samples; and a number of other products that either go to individuals or come from individuals. There is even a special class of such materials identified at § 173.6.

1.1.1.3.2 Other Regulated Materials–Domestic (ORM-D)

In the United States alone there is a very large class of materials that are shipped in such large volumes and are used in such great quantities by individuals at the retail or private consumer level as well as by smaller or service businesses that they are classified as "ORM-D," or Other Regulated Materials–Domestic. Materials that may be classified in this manner are regulated and have to meet some but not all of the regulatory requirements for shipment, storage, and handling incident to transportation; however, there is no requirement to identify these materials in their retail outlet, so that the consuming public at large, which can include smaller businesses and sole proprietorships, may be completely unaware that these materials are regulated as hazardous. There is tremendous amount of science to support a government position that allows such classification of materials. The problem arises because the general population does not understand the very real hazards associated with these materials, including their transportation as well as their use.

1.1.1.3.3 "Materials of Trade"

In 49 CFR at § 173.6, these materials are defined in excruciating but not necessarily clear detail. The bulk of these materials are not regulated in the eyes of the general public, and are not clearly labeled as hazardous materials within the retail setting. However, some require compliance with all portions of the regulations, while others are allowed to be handled with less than full adherence to the bulk of the regulatory requirements. Many such materials are found in retail outlets, in industrial or business settings, and are of concern during movement; therefore, they are "regulated." All of these would have MSDS sheets available. Some excellent examples would be cans of WD-40, windshield deicing cans, pump containers of fertilizer or insecticides, PVC cement, or fire extinguishers. You can buy all of these items in many retail outlets, and you will never see them marked or labeled as "hazardous materials" although there may be various warning labels on the individual containers or packaging. If you buy any of these items and you take them home, they are not subject to regulation in your vehicle or in your house. Unfortunately, in many jurisdictions they are not subject to any form of regulation for disposal either. On the other hand, if you are employed or self-employed and you carry these in a vehicle that is being used for business or in fact use them in the performance of your job, then they are subject to regulation.

1.1.1.3.4 Materials That Are Not Specifically Defined or Identified by Any Regulation but Represent a Hazard Strictly Because of Their Volume/Quantity or Other Distinguishing Characteristic

Every above-ground municipal water tank can legitimately be described as containing a hazardous material. The catastrophic failure of such water towers represents a potential flood that can put people and property at risk. During the preparation of this text, there was just such an event in Rochester, New Hampshire. In that case, there was not a total and complete catastrophic failure of a water tower but a significant breach. The response was exactly the same as for a toxic inhalation leak at the industrial facility: very large numbers of emergency personnel were activated, people were evacuated, and traffic diverted. In the end, the cost of that structural failure, regardless of liability and legal issues, will be borne by the city government and eventually passed on to its citizens through tax collections. In addition, many people were put at risk and were displaced, although in this case most of that displacement did not extend beyond 24 hours. If you want to learn more about this concept and the devastation such incidents can wreak, I suggest you take the time to learn a little about the "Boston Molasses Disaster" that occurred on January 15, 1919. In addition to the above-ground tank failure, there was significant structural damage, 21 people lost their lives, and a number of animals perished due to the unplanned and violent release of 2.3 million gallons of molasses into the streets of Boston. The case that has a certain amount of bitter irony to it would be the "London Beer Flood" of October 17, 1814. Only eight people are reported to have lost their lives, but the property damage immediately around the ruptured tanks was significant. We will go on to talk about this again later, but the release of a large quantity of any liquid being transported into any body of water is by definition an ecological disaster, and the material released, even if it is something as benign as saltwater or milk, can have a devastating effect on all living organisms that live in or on, or depend upon, that body of water. You cannot lose sight of this concept if you are involved in the operation of any large liquid processing facility or are responsible for large quantities of any liquids, whether in a static location or in transit.

1.2 The Difference Between a Consensus Standard and a Statutory Standard

Before one can understand, apply, and select standards, one must understand two or three key concepts. These are both legal and practical concepts that are keys to effective management of hazardous materials and are keys to effective compliance with statutory requirements.

A statutory standard is a legally binding requirement in that the requirement may be a requirement within the country of origin or internationally. Such standards often carry civil and criminal liabilities. Consensus standards are quite different although their impact on commercial operations and private individuals may be just as great. A consensus standard does not carry the weight of law and would ordinarily not allow for pursuit of a criminal case against the alleged wrongdoer. The most important standard or regulation for this text is, in fact, the equivalent of a consensus standard. The UN Orange Book, which we will discuss in greater detail early in the next chapter, does not have the weight of law. The UN itself does not have the legal standing to create enforceable regulations in this arena. There are numerous standards-setting organizations within the United States including such groups as the Association of American Railroads (AAR), the American Conference of Governmental Industrial Hygienists (ACGIH), the American National Standards Institute (ANSI), American Society for Testing and Materials (ASTM), the Building Officials Code Administrators International (BOCA), the Compressed Gas Association (CGA), the International Air Transportation Association (IATA), and the National Fire Protection Association (NFPA).

As shown in Appendix A7.2, a consensus standard may be included by reference within a federal code and therefore may be enforced as a statutory requirement. There must be specific reference to such a standard. Here is the specific language used in 49 CFR:

> *(a) Matter incorporated by reference—(1) General. There is incorporated, by reference in parts 170–189 of this subchapter, matter referred to that is not specifically set forth. This matter is hereby made a part of the regulations in parts 170–189 of this subchapter. The matter subject to change is incorporated only as it is in effect on the date of issuance of the regulation referring to that matter. The material listed in paragraph (a)(3) has been approved for incorporation by reference by the Director of the Federal Register in accordance with 5 U.S.C 552(a) and 1 CFR part 51. Material is incorporated as it exists on the date of the approval and a notice of any change in the material will be published in the Federal Register. Matters referenced by footnote are included as part of the regulations of this subchapter.*

With that foundation firmly established, it is time to look at the basic system used globally for the movement of goods that represent hazards to the global ecosystem, which means identifying and understanding the basic hazard classes and divisions used to move all such goods throughout the world.

1.3 The International Transportation Hazard Classes

There are nine basic hazard classes that are recognized internationally. Some of them are further broken down into divisions for a total of 20 universally used classes and divisions. In the United States there are additional classifications used, but they are not generally recognized anywhere else. Each of the most commonly used regulations provides definitions for the hazard classes. Table 1.1 allows you to identify not only where the definition appears within the appropriate regulation but also demonstrates the amount of harmonization standardization that has evolved since approximately 2000.

The following definitions are extracted from the 49 CFR; at the current time, they accurately reflect transportation definitions for the classes and divisions that are applicable in the United States.

§ 173.50 Class 1—Definitions

(a) *Explosive.* For the purposes of this subchapter, an *explosive* means any substance or article, including a device, which is designed to function by explosion (i.e., an extremely rapid release of gas and heat) or which, by chemical reaction within itself, is able to function in a similar manner even if not designed to function by explosion, unless the substance or article is otherwise classed under the provisions of this subchapter. The term includes a pyrotechnic substance or article, unless the substance or article is otherwise classed under the provisions of this subchapter.

(b) Explosives in Class 1 are divided into six divisions as follows:

(1) *Division 1.1* consists of explosives that have a mass explosion hazard. A mass explosion is one which affects almost the entire load instantaneously.

(2) *Division 1.2* consists of explosives that have a projection hazard but not a mass explosion hazard.

Table 1.1 The Nine Classes and Their Associated Divisions

Primary Class	*Divisions*	*49 CFR Definition*	*UN Definition*	*ICAO Definition*	*IMO Definition*	*ADR Definition*
1	1.1 EXPLOSIVE 1.1A 1	§ 173.50(b)1	2.1.1.4(a)	2.1.3(a)	2.1.1.4	2.2.1.1.5
	1.2 EXPLOSIVES 1.2B 1	§ 173.50(b)2	2.1.1.4(b)	2.1.3(b)	2.1.1.4	2.2.1.1.5
	1.3 EXPLOSIVES 1.3C 1	§ 173.50(b)3	2.1.1.4(c)	2.1.3(c)	2.1.1.4	2.2.1.1.5

	1.4	§ 173.50(b)4	2.1.1.4(d)	2.1.3(d)	2.1.1.4	2.2.1.1.5
	1.5	§ 173.50(b)5	2.1.1.4(e)	2.1.3(e)	2.1.1.4	2.2.1.1.5
	1.6	§ 173.50(b)6	2.1.1.4(f)	2.1.3(f)	2.1.1.4	2.2.1.1.5

Note: **See color insert.**

Table 1.1 ***(Continued)*** **The Nine Classes and Their Associated Divisions**

Primary Class	*Divisions*	*49 CFR Definition*	*UN Definition*	*ICAO Definition*	*IMO Definition*	*ADR Definition*
2	2.1 FLAMMABLE GAS 2	§ 173.115(a)	2.2.2.1(a)	2.2.1(a)	2.2.2.1	2.2.2.1.3
	2.2 NON-FLAMMABLE GAS 2	§ 173.115(b)	2.2.2.1(b)	2.2.1(b)	2.2.2.1	2.2.2.1.3
	2.2 OXYGEN 2	§ 173.115(b)	2.2.2.1(b)	2.2.1(b)	2.2.2.2	2.2.2.1.3

	TOXIC GAS 2 2.3 INHALATION HAZARD 2  	§ 173.115(c)	2.2.2.1(c)	2.2.1(c)	2.2.2.3	2.2.2.1.3
3[1]	 FLAMMABLE LIQUID 3	§ 173.120(a)	2.3.1	2.3.1.2	2.3.1	2.2.3.1.1

Note: **See color insert.**

Table 1.1 *(Continued)* The Nine Classes and Their Associated Divisions

Primary Class	*Divisions*	*49 CFR Definition*	*UN Definition*	*ICAO Definition*	*IMO Definition*	*ADR Definition*
4	4.1	§ 173.124(a)	2.4.1.1(a)	2.4.1.1(a)	2.4.2	2.2.41.1
	4.2	§ 173.124(b)	2.4.1.1(b)	2.4.1.1(b)	2.4.3	2.2.42.1.1
	4.3	§ 173.124(c)	2.4.1.1(c)	2.4.1.1(c)	2.4.4	2.2.43.1.1

5	5.1 OXIDIZER 5.1	§ 173.127(a)	2.5.1(a)	2.5.1(a)	2.5.2	2.2.51.1.1
	5.2 ORGANIC PEROXIDE 5.2	§ 173.128(a)	2.5.1(b)	2.5.1(b)	2.5.3	2.2.52.1.1

Note: **See color insert.**

Table 1.1 *(Continued)* The Nine Classes and Their Associated Divisions

Primary Class	*Divisions*	*49 CFR Definition*	*UN Definition*	*ICAO Definition*	*IMO Definition*	*ADR Definition*
6	6.1 TOXIC 6 / INHALATION HAZARD 6	§ 173.132(a)	2.6.1(a)	2.6.1(a)	2.6.2	2.2.61.1.1
	6.2 INFECTIOUS SUBSTANCE 6	§ 173.134(a)	2.6.1(b)	2.6.1(b)	2.6.3	2.2.62.1.1

7	RADIOACTIVE III CONTENTS ACTIVITY 7 RADIOACTIVE II CONTENTS ACTIVITY 7 RADIOACTIVE I CONTENTS ACTIVITY 7	§ 173.403	2.7.1.1	2.7.1.1	2.7.1	2.2.7.1.1

Note: **See color insert.**

Table 1.1 *(Continued)* The Nine Classes and Their Associated Divisions

Primary Class	*Divisions*	*49 CFR Definition*	*UN Definition*	*ICAO Definition*	*IMO Definition*	*ADR Definition*
8	CORROSIVE 8	§ 173.136(a)	2.8.1	2.8.1	2.8.1	2.2.8.1.1
9	9	§ 173.140(a)	2.9.1.1	2.9.1.1	2.9.1	2.2.9.1.1
ORM[2]		§ 173.144	N/A	N/A	N/A	

Note: **See color insert.**

(3) *Division 1.3* consists of explosives that have a fire hazard and either a minor blast hazard or a minor projection hazard or both, but not a mass explosion hazard.

(4) *Division 1.4* consists of explosives that present a minor explosion hazard. The explosive effects are largely confined to the package and no projection of fragments of appreciable size or range is to be expected. An external fire must not cause virtually instantaneous explosion of almost the entire contents of the package.

(5) *Division 1.5** consists of very insensitive explosives. This division is comprised of substances which have a mass explosion hazard but are so insensitive that there is very little probability of initiation or of transition from burning to detonation under normal conditions of transport.

* The probability of transition from burning to detonation is greater when large quantities are transported in a vessel.

(6) *Division 1.6** consists of extremely insensitive articles which do not have a mass explosive hazard. This division is comprised of articles which contain only extremely insensitive detonating substances and which demonstrate a negligible probability of accidental initiation or propagation.

§ 173.115 Class 2, Divisions 2.1, 2.2, and 2.3—Definitions

(a) *Division 2.1 (Flammable gas).* For the purpose of this subchapter, a *flammable gas* (Division 2.1) means any material which is a gas at 20°C (68°F) or less and 101.3 kPa (14.7 psia) of pressure (a material which has a boiling point of 20°C (68°F) or less at 101.3 kPa (14.7 psia)) which—
  (1) Is ignitable at 101.3 kPa (14.7 psia) when in a mixture of 13% or less by volume with air; or
  (2) Has a flammable range at 101.3 kPa (14.7 psia) with air of at least 12% regardless of the lower limit. Except for aerosols, the limits specified in paragraphs (a)(1) and (a)(2) of this section shall be determined at 101.3 kPa (14.7 psia) of pressure and a temperature of 20°C (68°F) in accordance with the ASTM E681–85, Standard Test Method for Concentration Limits of Flammability of Chemicals or other equivalent method approved by the Associate Administrator. The flammability of aerosols is determined by the tests specified in § 173.115 (k) of this section.

* The risk from articles of Division 1.6 is limited to the explosion of a single article.

(b) *Division 2.2 (non-flammable, nonpoisonous compressed gas—including compressed gas, liquefied gas, pressurized cryogenic gas, compressed gas in solution, asphyxiant gas and oxidizing gas).* For the purpose of this subchapter, a non-flammable, nonpoisonous compressed gas (Division 2.2) means any material (or mixture) which—
  (1) Exerts in the packaging a gauge pressure of 200 kPa (29.0 psig/43.8 psia) or greater at 20°C (68°F), is a liquefied gas or is a cryogenic liquid, and
  (2) Does not meet the definition of Division 2.1 or 2.3.

(c) *Division 2.3 (Gas poisonous by inhalation).* For the purpose of this subchapter, a *gas poisonous by inhalation* (Division 2.3) means a material which is a gas at 20°C (68°F) or less and a pressure of 101.3 kPa (14.7 psia) (a material which has a boiling point of 20°C (68°F) or less at 101.3 kPa (14.7 psia)) and which—
  (1) Is known to be so toxic to humans as to pose a hazard to health during transportation, or
  (2) In the absence of adequate data on human toxicity, is presumed to be toxic to humans because when tested on laboratory animals it has an LC_{50} value of not more than 5000 mL/m^3 (see § 173.116(d) of this subpart for assignment of Hazard Zones A, B, C, or D).

§ 173.120 Class 3—Definitions

(a) *Flammable liquid.* For the purpose of this subchapter, a *flammable liquid* (Class 3) means a liquid having a flash point of not more than 60°C (140°F), or any material in a liquid phase with a flash point at or above 37.8°C (100°F) that is intentionally heated and offered for transportation or transported at or above its flash point in a bulk packaging, with the following exceptions:
   (1) Any liquid meeting one of the definitions specified in § 173.115.
   (2) Any mixture having one or more components with a flash point of 60°C (140°F) or higher, that make up at least 99% of the total volume of the mixture, if the mixture is not offered for transportation or transported at or above its flash point.
   (3) Any liquid with a flash point greater than 35°C (95°F) that does not sustain combustion according to ASTM D 4206 (IBR, see § 171.7 of this subchapter) or the procedure in Appendix H of this part.
   (4) Any liquid with a flash point greater than 35°C (95°F) and with a fire point greater than 100°C (212°F) according to ISO 2592 (IBR, see § 171.7 of this subchapter).
   (5) Any liquid with a flash point greater than 35°C (95°F) which is in a water-miscible solution with a water content of more than 90% by mass.

(b) Combustible liquid.
   (1) For the purpose of this subchapter, a *combustible liquid* means any liquid that does not meet the definition of any other hazard class specified in this subchapter and has a flash point above 60°C (140°F) and below 93°C (200°F).
   (2) A flammable liquid with a flash point at or above 38°C (100°F) that does not meet the definition of any other hazard class may be reclassed as a combustible liquid. This provision does not apply to transportation by vessel or aircraft, except where other means of transportation is impracticable. An elevated temperature material that meets the definition of a Class 3 material because it is intentionally heated and offered for transportation or transported at or above its flash point may not be reclassed as a combustible liquid.
   (3) A combustible liquid that does not sustain combustion is not subject to the requirements of this subchapter as a combustible liquid. Either the test method specified in ASTM D 4206 or the procedure in Appendix H of this part may be used to determine if a material sustains combustion when heated under test conditions and exposed to an external source of flame.

(c) Flash point.
   (1) *Flash point* means the minimum temperature at which a liquid gives off vapor within a test vessel in sufficient concentration to form an ignitable mixture with air near the surface of the liquid. It shall be determined as follows:
      (i) For a homogeneous, single-phase, liquid having a viscosity less than 45 S.U.S. at 38°C (100°F) that does not form a surface film while under test, one of the following test procedures shall be used:
         (A) Standard Method of Test for Flash Point by Tag Closed Tester, (ASTM D 56);
         (B) Standard Methods of Test for Flash Point of Liquids by Setaflash Closed Tester, (ASTM D 3278); or
         (C) Standard Test Methods for Flash Point by Small Scale Closed Tester, (ASTM D 3828).

(ii) For a liquid other than one meeting all of the criteria of paragraph (c)(1)(i) of this section, one of the following test procedures shall be used:

(A) Standard Method of Test for Flash Point by Pensky—Martens Closed Tester, (ASTM D 93). For cutback asphalt, use Method B of ASTM D 93 or alternate tests authorized in this standard; or

(B) Standard Methods of Test for Flash Point of Liquids by Setaflash Closed Tester (ASTM D 3278).

(2) For a liquid that is a mixture of compounds that have different volatility and flash points, its flash point shall be determined as specified in paragraph (c)(1) of this section, on the material in the form in which it is to be shipped. If it is determined by this test that the flash point is higher than -7°C (20°F) a second test shall be made as follows: a portion of the mixture shall be placed in an open beaker (or similar container) of such dimensions that the height of the liquid can be adjusted so that the ratio of the volume of the liquid to the exposed surface area is 6 to one. The liquid shall be allowed to evaporate under ambient pressure and temperature (20 to 25°C [68 to 77°F]) for a period of 4 hours or until 10% by volume has evaporated, whichever comes first. A flash point is then run on a portion of the liquid remaining in the evaporation container and the lower of the two flash points shall be the flash point of the material.

(3) For flash point determinations by Setaflash closed tester, the glass syringe specified need not be used as the method of measurement of the test sample if a minimum quantity of 2 mL (0.1 ounce) is assured in the test cup.

(d) If experience or other data indicate that the hazard of a material is greater or less than indicated by the criteria specified in paragraphs (a) and (b) of this section, the Associate Administrator may revise the classification or make the material subject or not subject to the requirements of parts 170–189 of this subchapter.

§ 173.124 Class 4, Divisions 4.1, 4.2, and 4.3—Definitions

(a) *Division 4.1 (Flammable Solid).* For the purposes of this subchapter, *flammable solid* (Division 4.1) means any of the following three types of materials:

(1) Desensitized explosives that—

(i) When dry are Explosives of Class 1 other than those of compatibility group A, which are wetted with sufficient water, alcohol, or plasticizer to suppress explosive properties; and

(ii) Are specifically authorized by name either in the § 172.101 Table or have been assigned a shipping name and hazard class by the Associate Administrator under the provisions of—

(A) A special permit issued under subchapter A of this chapter; or
(B) An approval issued under § 173.56(i) of this part.

(2) (i) Self-reactive materials are materials that are thermally unstable and that can undergo a strongly exothermic decomposition even without participation of oxygen (air). A material is excluded from this definition if any of the following applies:
(A) The material meets the definition of an explosive as prescribed in subpart C of this part, in which case it must be classed as an explosive;
(B) The material is forbidden from being offered for transportation according to § 172.101 of this subchapter or § 173.21;
(C) The material meets the definition of an oxidizer or organic peroxide as prescribed in subpart D of this part, in which case it must be so classed;
(D) The material meets one of the following conditions:
(1) Its heat of decomposition is less than 300 J/g; or
(2) Its self-accelerating decomposition temperature (SADT) is greater than 75°C (167°F) for a 50 kg package; or
(3) It is an oxidizing substance in Division 5.1 containing less than 5.0% combustible organic substances; or
(E) The Associate Administrator has determined that the material does not present a hazard which is associated with a Division 4.1 material.

(ii) *Generic types.* Division 4.1 self-reactive materials are assigned to a generic system consisting of seven types. A self-reactive substance identified by technical name in the Self-Reactive Materials Table in § 173.224 is assigned to a generic type in accordance with that table. Self-reactive materials not identified in the Self-Reactive Materials Table in § 173.224 are assigned to generic types under the procedures of paragraph (a)(2)(iii) of this section.
(A) *Type A.* Self-reactive material type A is a self-reactive material which, as packaged for transportation, can detonate or deflagrate rapidly. Transportation of type A self-reactive material is forbidden.
(B) *Type B.* Self-reactive material type B is a self-reactive material which, as packaged for transportation, neither detonates nor deflagrates rapidly, but is liable to undergo a thermal explosion in a package.
(C) *Type C.* Self-reactive material type C is a self-reactive material which, as packaged for transportation, neither detonates nor deflagrates rapidly and cannot undergo a thermal explosion.
(D) *Type D.* Self-reactive material type D is a self-reactive material which—
(1) Detonates partially, does not deflagrate rapidly and shows no violent effect when heated under confinement;
(2) Does not detonate at all, deflagrates slowly and shows no violent effect when heated under confinement; or
(3) Does not detonate or deflagrate at all and shows a medium effect when heated under confinement.
(E) *Type E.* Self-reactive material type E is a self-reactive material which, in laboratory testing, neither detonates nor deflagrates at all and shows only a low or no effect when heated under confinement.

(F) *Type F.* Self-reactive material type F is a self-reactive material which, in laboratory testing, neither detonates in the cavitated state nor deflagrates at all and shows only a low or no effect when heated under confinement as well as low or no explosive power.

(G) *Type G.* Self-reactive material type G is a self-reactive material which, in laboratory testing, does not detonate in the cavitated state, will not deflagrate at all, shows no effect when heated under confinement, nor shows any explosive power. A type G self-reactive material is not subject to the requirements of this subchapter for self-reactive material of Division 4.1 provided that it is thermally stable (self-accelerating decomposition temperature is 50°C (122°F) or higher for a 50 kg (110 pounds) package). A self-reactive material meeting all characteristics of type G except thermal stability is classed as a type F self-reactive, temperature control material.

(iii) *Procedures for assigning a self-reactive material to a generic type.* A self-reactive material must be assigned to a generic type based on—

(A) Its physical state (i.e., liquid or solid), in accordance with the definition of liquid and solid in § 171.8 of this subchapter;

(B) A determination as to its control temperature and emergency temperature, if any, under the provisions of § 173.21(f);

(C) Performance of the self-reactive material under the test procedures specified in the UN Manual of Tests and Criteria (IBR, see § 171.7 of this subchapter) and the provisions of paragraph (a)(2)(iii) of this section; and

(D) Except for a self-reactive material which is identified by technical name in the Self-Reactive Materials Table in § 173.224(b) or a self-reactive material which may be shipped as a sample under the provisions of § 173.224, the self-reactive material is approved in writing by the Associate Administrator. The person requesting approval shall submit to the Associate Administrator the tentative shipping description and generic type and—

(1) All relevant data concerning physical state, temperature controls, and test results; or

(2) An approval issued for the self-reactive material by the competent authority of a foreign government.

(iv) *Tests.* The generic type for a self-reactive material must be determined using the testing protocol from Figure 14.2 (Flow Chart for Assigning Self-Reactive Substances to Division 4.1) from the UN Manual of Tests and Criteria.

(3) Readily combustible solids are materials that—

(i) Are solids which may cause a fire through friction, such as matches;

(ii) Show a burning rate faster than 2.2 mm (0.087 inches) per second when tested in accordance with the UN Manual of Tests and Criteria (IBR, see § 171.7 of this subchapter); or

(iii) Any metal powders that can be ignited and react over the whole length of a sample in 10 minutes or less, when tested in accordance with the UN Manual of Tests and Criteria.

(b) For the purposes of this subchapter, *spontaneously combustible material* (Division 4.2) means—
   (1) A pyrophoric material. A pyrophoric material is a liquid or solid that, even in small quantities and without an external ignition source, can ignite within five (5) minutes after coming in contact with air when tested according to UN Manual of Tests and Criteria.
   (2) A self-heating material. A self-heating material is a material that, when in contact with air and without an energy supply, is liable to self-heat. A material of this type which exhibits spontaneous ignition or if the temperature of the sample exceeds 200°C (392°F) during the 24-hour test period when tested in accordance with UN Manual of Tests and Criteria, is classed as a Division 4.2 material.

(c) *Division 4.3 (Dangerous when wet material).* For the purposes of this chapter, *dangerous when wet material* (Division 4.3) means a material that, by contact with water, is liable to become spontaneously flammable or to give off flammable or toxic gas at a rate greater than 1 L per kilogram of the material, per hour, when tested in accordance with UN Manual of Tests and Criteria.

§ 173.127 Class 5, Division 5.1—Definition and Assignment of Packing Groups

(a) *Definition.* For the purpose of this subchapter, *oxidizer* (Division 5.1) means a material that may, generally by yielding oxygen, cause or enhance the combustion of other materials.

(1) A solid material is classed as a Division 5.1 material if, when tested in accordance with the UN Manual of Tests and Criteria (IBR, see § 171.7 of this subchapter), its mean burning time is less than or equal to the burning time of a 3:7 potassium bromate/cellulose mixture.
(2) A liquid material is classed as a Division 5.1 material if, when tested in accordance with the UN Manual of Tests and Criteria, it spontaneously ignites or its mean time for a pressure rise from 690 kPa to 2070 kPa gauge is less than the time of a 1:1 nitric acid (65%)/cellulose mixture.

§ 173.128 Class 5, Division 5.2—Definitions and Types

(a) *Definitions.* For the purposes of this subchapter, *organic peroxide (Division 5.2)* means any organic compound containing oxygen (O) in the bivalent -O-O- structure and which may be considered a derivative of hydrogen peroxide, where one or more of the hydrogen atoms have been replaced by organic radicals, unless any of the following paragraphs applies:
  (1) The material meets the definition of an explosive as prescribed in subpart C of this part, in which case it must be classed as an explosive;
  (2) The material is forbidden from being offered for transportation according to § 172.101 of this subchapter or § 173.21;
  (3) The Associate Administrator has determined that the material does not present a hazard which is associated with a Division 5.2 material; or
  (4) The material meets one of the following conditions:
    (i) For materials containing no more than 1.0% hydrogen peroxide, the available oxygen, as calculated using the equation in paragraph (a)(4)(ii) of this section, is not more than 1.0%, or
    (ii) For materials containing more than 1.0% but not more than 7.0% hydrogen peroxide, the available oxygen, content (O_a) is not more than 0.5%, when determined using the equation:

$$O_a = 16 \times \sum_{i=1}^{k} \frac{n_i c_i}{m_i}$$

where, for a material containing k species of organic peroxides:
n_i = number of -O-O- groups per molecule of the *i* th species
c_i = concentration (mass %) of the *i* th species
m_i = molecular mass of the *i* th species

(b) *Generic types.* Division 5.2 organic peroxides are assigned to a generic system which consists of seven types. An organic peroxide identified by technical name in the Organic Peroxides Table in § 173.225 is assigned to a generic type in accordance with that table. Organic peroxides not identified in the Organic Peroxides table are assigned to generic types under the procedures of paragraph (c) of this section.

(1) *Type A.* Organic peroxide type A is an organic peroxide which can detonate or deflagrate rapidly as packaged for transport. Transportation of type A organic peroxides is forbidden.

(2) *Type B.* Organic peroxide type B is an organic peroxide which, as packaged for transport, neither detonates nor deflagrates rapidly, but can undergo a thermal explosion.

(3) *Type C.* Organic peroxide type C is an organic peroxide which, as packaged for transport, neither detonates nor deflagrates rapidly and cannot undergo a thermal explosion.

(4) *Type D.* Organic peroxide type D is an organic peroxide which—

(i) Detonates only partially, but does not deflagrate rapidly and is not affected by heat when confined;

(ii) Does not detonate, deflagrates slowly, and shows no violent effect if heated when confined; or

(iii) Does not detonate or deflagrate, and shows a medium effect when heated under confinement.

(5) *Type E.* Organic peroxide type E is an organic peroxide which neither detonates nor deflagrates and shows low, or no, effect when heated under confinement.

(6) *Type F.* Organic peroxide type F is an organic peroxide which will not detonate in a cavitated state, does not deflagrate, shows only a low, or no, effect if heated when confined, and has low, or no, explosive power.

(7) *Type G.* Organic peroxide type G is an organic peroxide which will not detonate in a cavitated state, will not deflagrate at all, shows no effect when heated under confinement, and shows no explosive power. A type G organic peroxide is not subject to the requirements of this subchapter for organic peroxides of Division 5.2 provided that it is thermally stable (self-accelerating decomposition temperature is 50°C (122°F) or higher for a 50 kg (110 pounds) package). An organic peroxide meeting all characteristics of type G except thermal stability and requiring temperature control is classed as a type F, temperature control organic peroxide.

(c) *Procedure for assigning an organic peroxide to a generic type.* An organic peroxide shall be assigned to a generic type based on—

(1) Its physical state (i.e., liquid or solid), in accordance with the definitions for liquid and solid in § 171.8 of this subchapter;

(2) A determination as to its control temperature and emergency temperature, if any, under the provisions of § 173.21(f); and

(3) Performance of the organic peroxide under the test procedures specified in the UN Manual of Tests and Criteria (IBR, see § 171.7 of this subchapter), and the provisions of paragraph (d) of this section.

(d) *Approvals.* (1) An organic peroxide must be approved, in writing, by the Associate Administrator, before being offered for transportation or transported, including assignment of a generic type and shipping description, except for—

(i) An organic peroxide which is identified by technical name in the Organic Peroxides Table in § 173.225(c);

(ii) A mixture of organic peroxides prepared according to § 173.225(b); or

(iii) An organic peroxide which may be shipped as a sample under the provisions of § 173.225(b).

(2) A person applying for an approval must submit all relevant data concerning physical state, temperature controls, and tests results or an approval issued for the organic peroxide by the competent authority of a foreign government.

(e) *Tests.* The generic type for an organic peroxide shall be determined using the testing protocol from Figure 20.1(a) (Classification and Flow Chart Scheme for Organic Peroxides) from the UN Manual of Tests and Criteria (IBR, see § 171.7 of this subchapter).

§ 173.132 Class 6, Division 6.1—Definitions

(a) For the purpose of this subchapter, *poisonous material* (Division 6.1) means a material, other than a gas, which is known to be so toxic to humans as to afford a hazard to health during transportation, or which, in the absence of adequate data on human toxicity:

(1) Is presumed to be toxic to humans because it falls within any one of the following categories when tested on laboratory animals (whenever possible, animal test data that has been reported in the chemical literature should be used):

(i) *Oral Toxicity.* A liquid or solid with an LD_{50} for acute oral toxicity of not more than 300 mg/kg.

(ii) *Dermal Toxicity.* A material with an LD_{50} for acute dermal toxicity of not more than 1000 mg/kg.

(iii) *Inhalation Toxicity.* (A) A dust or mist with an LC_{50} for acute toxicity on inhalation of not more than 4 mg/L; or

(B) A material with a saturated vapor concentration in air at 20°C (68°F) greater than or equal to one-fifth of the LC_{50} for acute toxicity on inhalation of vapors and with an LC_{50} for acute toxicity on inhalation of vapors of not more than 5000 mL/m^3; or

(2) Is an irritating material, with properties similar to tear gas, which causes extreme irritation, especially in confined spaces.

(b) For the purposes of this subchapter—

(1) LD_{50} (median lethal dose) for acute oral toxicity is the statistically derived single dose of a substance that can be expected to cause death within 14 days in 50% of young adult albino rats when administered by the oral route. The LD_{50}value is expressed in terms of mass of test substance per mass of test animal (mg/kg).

(2) LD_{50} for acute dermal toxicity means that dose of the material which, administered by continuous contact for 24 hours with the shaved intact skin (avoiding abrading) of an albino rabbit, causes death within 14 days in half of the animals tested. The number of animals tested must be sufficient to give statistically valid results and be in conformity with good pharmacological practices. The result is expressed in mg/kg body mass.

(3) LC_{50} for acute toxicity on inhalation means that concentration of vapor, mist, or dust which, administered by continuous inhalation for one hour to both male and female young

adult albino rats, causes death within 14 days in half of the animals tested. If the material is administered to the animals as a dust or mist, more than 90% of the particles available for inhalation in the test must have a diameter of 10 microns or less if it is reasonably foreseeable that such concentrations could be encountered by a human during transport. The result is expressed in mg/L of air for dusts and mists or in mL/m^3 of air (parts per million) for vapors. See § 173.133(b) for LC_{50}determination for mixtures and for limit tests.

(i) When provisions of this subchapter require the use of the LC_{50} for acute toxicity on inhalation of dusts and mists based on a one-hour exposure and such data is not available, the LC_{50} for acute toxicity on inhalation based on a four-hour exposure may be multiplied by four and the product substituted for the one-hour LC_{50} for acute toxicity on inhalation.

(ii) When the provisions of this subchapter require the use of the LC_{50} for acute toxicity on inhalation of vapors based on a one-hour exposure and such data is not available, the LC_{50} for acute toxicity on inhalation based on a four-hour exposure may be multiplied by two and the product substituted for the one-hour LC_{50} for acute toxicity on inhalation.

(iii) A solid substance should be tested if at least 10% of its total mass is likely to be dust in a respirable range, e.g., the aerodynamic diameter of that particle-fraction is 10 microns or less. A liquid substance should be tested if a mist is likely to be generated in a leakage of the transport containment. In carrying out the test both for solid and liquid substances, more than 90% (by mass) of a specimen prepared for inhalation toxicity testing must be in the respirable range as defined in this paragraph (b)(3)(iii).

(c) For purposes of classifying and assigning packing groups to mixtures possessing oral or dermal toxicity hazards according to the criteria in § 173.133(a)(1), it is necessary to determine the acute LD_{50} of the mixture. If a mixture contains more than one active constituent, one of the following methods may be used to determine the oral or dermal LD_{50} of the mixture:

(1) Obtain reliable acute oral and dermal toxicity data on the actual mixture to be transported;

(2) If reliable, accurate data is not available, classify the formulation according to the most hazardous constituent of the mixture as if that constituent were present in the same concentration as the total concentration of all active constituents; or

(3) If reliable, accurate data is not available, apply the formula:

$$\frac{C_{A+}}{T_A} + \frac{C_B}{T_B} + \frac{C_Z}{T_Z} = \frac{100}{T_M}$$

where:

C = the % concentration of constituent A, B ... Z in the mixture;

T = the oral LD_{50}values of constituent A, B ... Z;

T_M = the oral LD_{50}value of the mixture.

Note to formula in paragraph (c)(3): This formula also may be used for dermal toxicities provided that this information is available on the same species for all constituents. The use of this formula does not take into account any potentiation or protective phenomena.

(d) The foregoing categories shall not apply if the Associate Administrator has determined that the physical characteristics of the material or its probable hazards to humans as shown by documented experience indicate that the material will not cause serious sickness or death.

§ 173.134 Class 6, Division 6.2—Definitions and Exceptions

(a) *Definitions and classification criteria.* For the purposes of this subchapter, the following definitions and classification criteria apply to Division 6.2 materials.

(1) *Division 6.2 (Infectious substance)* means a material known or reasonably expected to contain a pathogen. A pathogen is a microorganism (including bacteria, viruses, rickettsiae, parasites, fungi) or other agent, such as a proteinaceous infectious particle (prion), that can cause disease in humans or animals. An infectious substance must be assigned the identification number UN 2814, UN 2900, UN 3373, or UN 3291 as appropriate, and must be assigned to one of the following categories:

(i) *Category A:* An infectious substance in a form capable of causing permanent disability or life-threatening or fatal disease in otherwise healthy humans or animals when exposure to it occurs. An exposure occurs when an infectious substance is released outside of its protective packaging, resulting in physical contact with humans or animals. A Category A infectious substance must be assigned to identification number UN 2814 or UN 2900, as appropriate. Assignment to UN 2814 or UN 2900 must be based on the known medical history or symptoms of the source patient or animal, endemic local conditions, or professional judgment concerning the individual circumstances of the source human or animal.

(ii) *Category B:* An infectious substance that is not in a form generally capable of causing permanent disability or life-threatening or fatal disease in otherwise healthy humans or animals when exposure to it occurs. This includes Category B infectious substances transported for diagnostic or investigational purposes. A Category B infectious substance must be described as "Biological substance, Category B" and assigned identification number UN 3373. This does not include regulated medical waste, which must be assigned identification number UN 3291.

(2) *Biological product* means a virus, therapeutic serum, toxin, antitoxin, vaccine, blood, blood component or derivative, allergenic product, or analogous product, or arsphenamine or derivative of arsphenamine (or any other trivalent arsenic compound) applicable to the prevention, treatment, or cure of a disease or condition of human beings or animals. A *biological product* includes a material subject to regulation under 42 U.S.C. 262 or 21 U.S.C. 151–159. Unless otherwise excepted, a *biological product* known or reasonably expected to contain a pathogen that meets the definition of a Category A or B infectious substance must be assigned the identification number UN 2814, UN 2900, or UN 3373, as appropriate.

(3) *Culture* means an infectious substance containing a pathogen that is intentionally propagated. *Culture* does not include a human or animal patient specimen as defined in paragraph (a)(4) of this section.

(4) *Patient specimen* means human or animal material collected directly from humans or animals and transported for research, diagnosis, investigational activities, or disease treatment or prevention. *Patient specimen* includes excreta, secreta, blood and its components, tissue and tissue swabs, body parts, and specimens in transport media (e.g., transwabs, culture media, and blood culture bottles).

(5) *Regulated medical waste or clinical waste or (bio) medical waste* means a waste or reusable material derived from the medical treatment of an animal or human, which includes diagnosis and immunization, or from biomedical research, which includes the production and testing of biological products. Regulated medical waste or clinical waste or (bio) medical waste containing a Category A infectious substance must be classed as an infectious substance, and assigned to UN 2814 or UN 2900, as appropriate.

(6) *Sharps* means any object contaminated with a pathogen or that may become contaminated with a pathogen through handling or during transportation and also capable of cutting or penetrating skin or a packaging material. *Sharps* includes needles, syringes, scalpels, broken glass, culture slides, culture dishes, broken capillary tubes, broken rigid plastic, and exposed ends of dental wires.

(7) *Toxin* means a Division 6.1 material from a plant, animal, or bacterial source. A *toxin* containing an infectious substance or a *toxin* contained in an infectious substance must be classed as Division 6.2, described as an infectious substance, and assigned to UN 2814 or UN 2900, as appropriate.

(8) *Used health care product* means a medical, diagnostic, or research device or piece of equipment, or a personal care product used by consumers, medical professionals, or pharmaceutical providers that does not meet the definition of a patient specimen, biological product, or regulated medical waste, is contaminated with potentially infectious body fluids or materials, and is not decontaminated or disinfected to remove or mitigate the infectious hazard prior to transportation.

(b) *Exceptions.* The following are not subject to the requirements of this subchapter as Division 6.2 materials:

(1) A material that does not contain an infectious substance or that is unlikely to cause disease in humans or animals.

(2) Non-infectious biological materials from humans, animals, or plants. Examples include non-infectious cells, tissue cultures, blood or plasma from individuals not suspected of having an infectious disease, DNA, RNA or other non-infectious genetic elements.

(3) A material containing micro-organisms that are non-pathogenic to humans or animals.

(4) A material containing pathogens that have been neutralized or inactivated such that they no longer pose a health risk.

(5) A material with a low probability of containing an infectious substance, or where the concentration of the infectious substance is at a level naturally occurring in the environment so it cannot cause disease when exposure to it occurs. Examples of these materials include: Foodstuffs; environmental samples, such as water or a sample of dust or mold; and substances that have been treated so that the pathogens have been neutralized or deactivated, such as a material treated by steam sterilization, chemical disinfection, or other appropriate method, so it no longer meets the definition of an infectious substance.

(6) A biological product, including an experimental or investigational product or component of a product, subject to Federal approval, permit, review, or licensing requirements, such as those required by the Food and Drug Administration of the U.S. Department of Health and Human Services or the U.S. Department of Agriculture.

(7) Blood collected for the purpose of blood transfusion or the preparation of blood products; blood products; plasma; plasma derivatives; blood components; tissues or organs intended for use in transplant operations; and human cell, tissues, and cellular and tissue-based products regulated under authority of the Public Health Service Act (42 U.S.C. 264–272) and/or the Food, Drug, and Cosmetic Act (21 U.S.C. 332 *et seq.).*

(8) Blood, blood plasma, and blood components collected for the purpose of blood transfusion or the preparation of blood products and sent for testing as part of the collection process, except where the person collecting the blood has reason to believe it contains an infectious substance, in which case the test sample must be shipped as a Category A or Category B infectious substance in accordance with § 173.196 or § 173.199, as appropriate.

(9) Dried blood spots or specimens for fecal occult blood detection placed on absorbent filter paper or other material.

(10) A Division 6.2 material, other than a Category A infectious substance, contained in a patient sample being transported for research, diagnosis, investigational activities, or disease treatment or prevention, or a biological product, when such materials are transported by a private or contract carrier in a motor vehicle used exclusively to transport such materials. Medical or clinical equipment and laboratory products may be transported aboard the same vehicle provided they are properly packaged and secured against exposure or contamination. If the human or animal sample or biological product meets the definition of regulated medical waste in paragraph (a)(5) of this section, it must be offered for transportation and transported in conformance with the appropriate requirements for regulated medical waste.

(11) A human or animal sample (including, but not limited to, secreta, excreta, blood and its components, tissue and tissue fluids, and body parts) being transported for routine testing not related to the diagnosis of an infectious disease, such as for drug/alcohol testing, cholesterol testing, blood glucose level testing, prostate specific antibody testing, testing to monitor kidney or liver function, or pregnancy testing, or for tests for diagnosis of non-infectious diseases, such as cancer biopsies, and for which there is a low probability the sample is infectious.

(12) Laundry and medical equipment and used health care products, as follows:

(i) Laundry or medical equipment conforming to the regulations of the Occupational Safety and Health Administration of the Department of Labor in 29 CFR 1910.1030. This exception includes medical equipment intended for use, cleaning, or refurbishment, such as reusable surgical equipment, or equipment used for testing where the components within which the equipment is contained essentially function as packaging. This exception does not apply to medical equipment being transported for disposal.

(ii) Used health care products not conforming to the requirements in 29 CFR 1910.1030 and being returned to the manufacturer or the manufacturer's designee are excepted from the requirements of this subchapter when offered for transportation or transported in accordance with this paragraph (b)(12). For purposes of this paragraph, a health care product is used when it has been removed from its original packaging. Used health care products contaminated with or suspected of contamination with a Category A infectious substance may not be transported under the provisions of this paragraph.

(A) Each used health care product must be drained of free liquid to the extent practicable and placed in a watertight primary container designed and constructed to assure that it remains intact under conditions normally incident to transportation. For a used health care product capable of cutting or penetrating skin or packaging material, the primary container must be capable of retaining the product without puncture of the packaging under normal conditions of transport. Each primary container must be marked with a BIOHAZARD marking conforming to 29 CFR 1910.1030(g)(1)(i).

(B) Each primary container must be placed inside a watertight secondary container designed and constructed to assure that it remains intact under conditions normally incident to transportation. The secondary container must be marked with a BIOHAZARD marking conforming to 29 CFR 1910.1030(g)(1)(i).

(C) The secondary container must be placed inside an outer packaging with sufficient cushioning material to prevent movement between the secondary container and the outer packaging. An itemized list of the contents of the primary container and information concerning possible contamination with a Division 6.2 material, including its possible location on the product, must be placed between the secondary container and the outside packaging.

(D) Each person who offers or transports a used health care product under the provisions of this paragraph must know about the requirements of this paragraph.

(13) Any waste or recyclable material, other than regulated medical waste, including—
   (i) Household waste as defined in § 171.8, when transported in accordance with applicable state, local, or tribal requirements.
   (ii) Sanitary waste or sewage;
   (iii) Sewage sludge or compost;
   (iv) Animal waste generated in animal husbandry or food production; or
   (v) Medical waste generated from households and transported in accordance with applicable state, local, or tribal requirements.

(14) Corpses, remains, and anatomical parts intended for interment, cremation, or medical research at a college, hospital, or laboratory.

(15) Forensic material transported on behalf of a U.S. Government, state, local or Indian tribal government agency, except that—
   (i) Forensic material known or suspected to contain a Category B infectious substance must be shipped in a packaging conforming to the provisions of § 173.24.
   (ii) Forensic material known or suspected to contain a Category A infectious substance or an infectious substance listed as a select agent in 42 CFR Part 73 must be transported in packaging capable of meeting the test standards in § 178.609 of this subchapter. The secondary packaging must be marked with a BIOHAZARD symbol conforming to specifications in 29 CFR 1910.1030(g)(1)(i). An itemized list of contents must be enclosed between the secondary packaging and the outer packaging.

(16) Agricultural products and food as defined in the Federal Food, Drug, and Cosmetics Act (21 U.S.C. 332 *et seq.).*

(c) *Exceptions for regulated medical waste.* The following provisions apply to the transportation of regulated medical waste:
   (1) A regulated medical waste transported by a private or contract carrier is excepted from—
      (i) The requirement for an "INFECTIOUS SUBSTANCE" label if the outer packaging is marked with a "BIOHAZARD" marking in accordance with 29 CFR 1910.1030; and
      (ii) The specific packaging requirements of § 173.197, if packaged in a rigid non-bulk packaging conforming to the general packaging requirements of §§ 173.24 and

173.24a and packaging requirements specified in 29 CFR 1910.1030, provided the material does not include a waste concentrated stock culture of an infectious substance. Sharps containers must be securely closed to prevent leaks or punctures.

(2) A waste stock or culture of a Category B infectious substance may be offered for transportation and transported as a regulated medical waste when it is packaged in a rigid non-bulk packaging conforming to the general packaging requirements of §§ 173.24 and 173.24a and packaging requirements specified in 29 CFR 1910.1030 and transported by a private or contract carrier in a vehicle used exclusively to transport regulated medical waste. Medical or clinical equipment and laboratory products may be transported aboard the same vehicle provided they are properly packaged and secured against exposure or contamination. Sharps containers must be securely closed to prevent leaks or punctures.

(d) If an item listed in paragraph (b) or (c) of this section meets the definition of another hazard class or if it is a hazardous substance, hazardous waste, or marine pollutant, it must be offered for transportation and transported in accordance with applicable requirements of this subchapter.

§ 173.403 Class 7—Definitions

For purposes of this subchapter—

Radioactive material means any material containing radionuclides where both the activity concentration and the total activity in the consignment exceed the values specified in the table in § 173.436 or values derived according to the instructions in § 173.433.

§ 173.136 Class 8—Definitions

(a) For the purpose of this subchapter, "corrosive material" (Class 8) means a liquid or solid that causes full thickness destruction of human skin at the site of contact within a specified period of time. A liquid, or a solid which may become liquid during transportation, that has a severe corrosion rate on steel or aluminum based on the criteria in § 173.137(c)(2) is also a corrosive material.

(b) If human experience or other data indicate that the hazard of a material is greater or less than indicated by the results of the tests specified in paragraph (a) of this section, PHMSA may revise its classification or make the determination that the material is not subject to the requirements of this subchapter.
(c) Skin corrosion test data produced no later than September 30, 1995, using the procedures of part 173, appendix A, in effect on September 30, 1995 (see 49 CFR part 173, appendix A, revised as of October 1, 1994) for appropriate exposure times may be used for classification and assignment of packing group for Class 8 materials corrosive to skin.

§ 173.140 Class 9—Definitions

For the purposes of this subchapter, *miscellaneous hazardous material* (Class 9) means a material which presents a hazard during transportation but which does not meet the definition of any other hazard class. This class includes:

(a) Any material which has an anesthetic, noxious, or other similar property which could cause extreme annoyance or discomfort to a flight crew member so as to prevent the correct performance of assigned duties; or
(b) Any material that meets the definition in § 171.8 of this subchapter for an elevated temperature material, a hazardous substance, a hazardous waste, or a marine pollutant.

§ 173.144 Other Regulated Materials (ORM)—Definitions

For the purpose of this subchapter, "ORM-D material" means a material such as a consumer commodity, which, although otherwise subject to the regulations of this subchapter, presents a limited hazard during transportation due to its form, quantity and packaging. It must be a material for which exceptions are provided in the § 172.101 table. Each ORM-D material and category of ORM-D material is listed in the § 172.101 table.
(49 CFR, 2011)

1.4 Additional Definitions

Definitions in the previous section come directly from the regulations and apply to the issues related to the regulations. There are additional groups or categories of definitions that are important to both the strategic management and the operational management of hazardous materials in regard to life cycle management. This section looks at general business definitions and includes some terms that are essentially more technical than science-based.

1.4.1 General Definitions

CHEMTREC (USA)—A 24-hour emergency contact point that can be used to obtain guidance and advice relating to chemical spills that occur on public highways and railroads. Chemical Transportation Emergency Center, Chemical Manufacturers Association.

CHLOREP (USA)—Chlorine Emergency Plan. Operated 24 hours a day, 7 days a week by the Chlorine Institute for any emergency involving chlorine.

Compatible—Materials that are able to exist in close and permanent association indefinitely.

Competent Authority—A national (Department of Transportation (DOT) and International Maritime Organization (IMO)) law for control or regulation of a particular aspect of the transportation of hazardous materials (dangerous goods). The Director, Office of Hazardous Materials Transportation (OHMT) of the Research and Special Programs Administration (RSPA) of DOT, is the United States Competent Authority for the purposes of Competent Authority under Title 49 CFR, Title 46 CFR, parts 64 and 146, and IMO regulations.

Consignee—The addressee to whom the item is shipped.

Container—Any portable device in which a material is stored, transported, disposed of, or otherwise handled (see Title 40 CFR 260.10(a)(9)).

Container, Intermodal, ISO—An article of transport equipment that meets the standards of the International Organization for Standardization (ISO) designed to facilitate and optimize the carriage of goods by one or more modes of transportation without intermediate handling of the contents and equipped with features permitting ready handling and transfer from one mode to another. Containers may be fully enclosed with one or more doors, open top, tank, refrigerated, open rack, gondola, flat track, and other designs. Included in this definition are modules or arrays that can be coupled to form an integral unit regardless of intention to move singly or in multiplex configuration.

Containerization—The use of transport containers (container express (CONEX), military-owned demountable containers (MILVANs), commercially or government-owned (or leased) shipping containers (SEAVANs), and roll on/roll off (RORO) trailers) to unitize cargo for transportation, supply, and storage. Containerization aids carriage of goods by one or more modes of transportation without the need for intermediate handling of the contents, and incorporates supply, security, packaging, storage, and transportation into the distribution system from source to user.

Dangerous When Wet—A label required for certain materials being shipped under US DOT, IATA/ICAO, and IMO regulations (see Title 49 CFR 172.423).

Disposal Drum/Recovery Drum—A nonprofessional reference to a drum used to overpack damaged or leaking containers of hazardous materials for shipment; the proper shipping name is Salvage Drum as cited in Title 49 CFR 173.3.

Distribution System—A complex of facilities, methods, patterns, and procedures designed to receive, store, maintain, distribute, and control the flow of material from the point of receipt into a military supply system to the point of issue at a post, camp, station, base, or equivalent.

Dredging—To remove earth or silt from the bottom of a body of water.

Dunnage—Any material (boards, planks, blocks, pneumatic pillows) used to support or secure supplies in storage or in transit.

Empty Packaging—As related to Title 49 CFR: 1.

1. The description on the shipping paper for a packaging containing the residue of a hazardous material may include the words "RESIDUE: Last Contained * * *" in association with the basic description of the hazardous material last contained in the packaging.
2. For a tank car containing the residue (as defined in Title 49 CFR 171.8) of a hazardous material, the requirements of Title 49 CFR 172.203(e)(3) and 174.25(c) apply.
3. If a packaging, including a tank car, contains a residue that is a hazardous substance, the description on the shipping papers must be prefaced with the phrase "RESIDUE: Last Contained * * *" and the letters "RQ" must be entered on the shipping paper either before or after the description.

Expiration Date—The date by which nonextendable items (type I) should be discarded as no longer suitable for issue or use.

Explosion Proof—A device that is designed to withstand internal explosions and prevent hot vapors or particles from exiting before being significantly cooled.

Flammable Material—Any solid, liquid, vapor, or gas that will ignite easily and burn rapidly. Flammable solids are of several types: (1) dusts or fine powders; (2) those that ignite spontaneously at low temperatures; (3) those in which internal heat is built up by microbial or other degradation activity; (4) films, fibers, and fabrics of low-ignition point materials.

Flashback—A recession of flame into an unwanted location.

Full Protective Clothing—Such units are typically recommended where high chemical gas, vapor, or fume concentrations in air may have a corrosive effect on exposed skin, and where the chemical in air may be readily absorbed through the skin to produce toxic effects. These suits are impervious to chemicals, offer full body protection, and include self-contained breathing apparatus (SCBA (EPA level B protection)).

Fully Encapsulating Suits—Full chemical protective suits that are impervious to chemicals, offer full body protection from chemicals and their vapors/fumes, and are to be used with SCBA (EPA level A protection).

Hazard Class—A category of hazard associated with an HM/HW that has been determined capable of posing an unreasonable risk to health, safety, and property when transported (see Title 49 CFR 171.8).

Hazardous Chemicals—Hazardous materials used in the workplace that are regulated under OSHA "right-to-know" regulations in Title 29 CFR 1910.1200.

IMDG Designation—Hazardous material identifier published in the International Maritime Dangerous Goods Code (IMDGC).

Incompatible Waste—A waste unsuitable for commingling with another waste or material, where the commingling might result in the following:

1. Extreme heat or pressure generation.
2. Fire.
3. Explosion or violent reaction.
4. Formation of substances that are shock-sensitive, friction-sensitive, or otherwise have the potential to react violently.
5. Formation of toxic dusts, mists, fumes, gases, or other chemicals.
6. Volatilization of ignitable or toxic chemicals due to heat generation in such a manner that the likelihood of contamination of groundwater or escape of the substances into the environment is increased.

Insulator—Any substance or mixture that has an extremely low dielectric constant, low thermal conductivity, or both.

Material Safety Data Sheet—An MSDS must be in English and include information regarding the specific identity of hazardous chemicals. Also includes information on health effects, first aid, chemical and physical properties, and emergency phone numbers. (Latest revision of Federal Standard 313.)

MILVAN—Military-owned demountable container, conforming to U.S. and international standards and operated in a centrally controlled fleet for movement of military cargo.

Overpack—Except when referenced to a packaging specified in part 178 of Title 49 CFR, means an enclosure used by a single consignor to provide protection or convenience in handling of a package or to consolidate two or more packages. "Overpack" does not include a freight container.

Package or Outside Package—A packaging plus its contents (see Title 49 CFR 171.8).

Packaging—The assembly of one or more containers and any other components necessary to assure compliance with minimum packaging requirements; includes containers (other than freight containers or overpacks), portable tanks cargo tanks, tank cars, and multi-unit tank car tanks (see Title 49 CFR 171.8)

Packing—See Packaging.

Pallets—A pallet is a low, portable platform, constructed of wood, metal, plastic, or fiberboard, built to specified dimensions, on which supplies are loaded, transported, or stored in units. Flat pallets are either single-faced or double-faced. Single-faced pallets have one platform with stringers underneath on which the weight of the load rests. Double-faced pallets have two platforms separated by stringers. Pallets may afford two- or four-way entry. The two-way entry pallet is so constructed that the forks of a forklift truck may be inserted from either the front or rear of the pallet. The four-way pallet is so constructed that the forks of a forklift truck may be inserted from any of the four sides. Flat pallets are constructed of either softwood or hardwood. Expendable pallets are four-way entry and are composed of either fiberboard, polystyrene, or a combination of these. A box pallet is constructed with a framework and cross members extending up from a pallet platform, the front side normally being left open for loading or unloading (see DOD 4145.19-M-1, Storage and Materials Handling). Aircraft (463L): Aluminum air cargo pallet, 88 inches by 108 inches or 54 inches by 88 inches, on which shipments are consolidated for movement by AMC.

Placard—A small plate placed on transportation vehicles to facilitate the identification of the shipment.

Poison Control Centers—A nationwide network of poison control centers has been set up with the aid of the U.S. Food and Drug Administration and Department of Health and Human Services. The centers, usually established in local hospitals, are now widely distributed and available by phone from most parts of the country. Staff members are specially trained in the treatment of poisoning cases.

Potential—Ability for something to occur.

Propagate—To cause to spread out and affect a greater area.

Proper Shipping Name—The name of the hazardous material shown in Roman print in Title 49 CFR 172.101.

Protective Clothing Required—Characteristics and OSHA criteria for required protection against contact with the material.

Recoupment—The process of regaining or recovering materials that are still useful through repackaging, repacking, remarking, or repairing. (Pesticides regulated by Title 40 CFR under FIFRA will not be eligible for recoupment or relabeling.)

Residue—As referenced in Title 49 CFR 171.8, residue is the hazardous material remaining in a packaging after its contents have been emptied and before the packaging is refilled, or cleaned and purged of vapor to remove any potential hazard. Residue of a hazardous material, as applied to the contents of a tank car (other than DOT specification 106 or 110 tank cars), is a quantity of material no greater than 3% of the car's marked volumetric capacity.

Rinse Liquid—The liquid(s) that remains after a container has been rinsed out.

Salvage Drum—A drum with a removable metal head that is compatible with the lading used to transport damaged or leaking hazardous materials for repackaging or disposal (see Title 49 CFR).

SEAVAN—Commercial or government-owned (or leased) shipping container.

Shelf Life—The total period of time beginning with the date of manufacture/cure/assembly/pack that an item may remain in the combined wholesale (including manufacturer) and retail storage system and still remain suitable for issue/use by the end user. Shelf life is not to be confused with service life, which is a measure of mean life of an item.

Shelf Life Item—An item of supply possessing deteriorative or unstable characteristics to the degree that a storage time period must be assigned to ensure that it will perform satisfactorily in service.

Static Electricity—Charges created by the conductive movements of unlike materials.

Storage Serviceability Standard—Mandatory instructions for the inspection, testing, and/or restoration of items in storage, encompassing storage criteria, preservation, packaging, packing, and marking requirements, and time phasing for inspection during the storage cycle to determine the material serviceability and the degree of degradation that has occurred. Storage Serviceability Standards must be prepared by the managing wholesale ICP or other responsible organization for type II shelf life items only. They are used at the wholesale and retail level to determine if type II shelf life items have retained sufficient quantities of their original characteristics and are of a quality level that warrants extension of their assigned time period, and the length of the time period extension.

Toxic—Any material that can cause harm by either ingestion, inhalation, or absorption via the skin.

Toxicity Categories—EPA has published regulations for use of human hazard signal words on pesticide labels. The following signal words are assigned by levels of toxicity:

(1) Toxicity Category I. All pesticide products meeting the criteria of Toxicity Category I shall bear on the front panel the signal word "Danger." In addition, if the product has been assigned to Toxicity Category I on the basis of its oral, inhalation, or dermal toxicity (as distinct from skin and eye local effects), the word "Poison" shall appear in red on a background of distinctly contrasting color, and the skull and crossbones shall appear in immediate proximity to the word "Poison."

(2) Toxicity Category II. All pesticide products meeting the criteria of Toxicity Category II shall bear on the front panel the signal word "Warning."

(3) Toxicity Category III. All pesticide products meeting the criteria of Toxicity Category III shall bear on the front panel the signal word "Caution."

(4) Toxicity Category IV. All pesticide products meeting the criteria of Toxicity Category IV shall bear on the front panel the signal word "Caution."

(5) Use of signal words. Use of signal word(s) associated with a higher toxicity category is not permitted except when the agency determines that such labeling is necessary to prohibit unacceptable adverse effects.

Type I Shelf Life Item—An individual item of supply determined through an evaluation of technical test data and/or actual experience to be an item with a definite nonextendable period of shelf life.

Type II Shelf Life Item—An individual item of supply having an assigned shelf life time period that may be extended after completion of inspection/test/restoration action.

This next set of definitions are science-based definitions rather than regulatory-based definitions, and many of these definitions and the concepts they represent are critical to the decision-making process from product concept and design to final disposal or reuse.

1.4.2 Scientific Terms and Definitions

Acidic—A sour substance; specifically any of various typically water-soluble and sour compounds that are capable of reacting with a base to form a salt, that redden litmus, that are hydrogen-containing molecules or ions able to give up a proton to a base, or that are substances able to accept an unshared pair of electrons from a base.

Aerosol—Solid or liquid particles, usually less than 1 micron in diameter, suspended in a gaseous medium.

Alkali—Any substance that in water solution is bitter, more or less irritating, or caustic to the skin. Strong alkalis in solution are corrosive to the skin and mucous membranes.

Ambient Air—The air of the environment in which an experiment is conducted or in which any physical or chemical event occurs.

Anhydrous—Free from water.

Auto-Ignition Temperature—The minimum temperature required to initiate or cause self-sustained combustion in any substance in the absence of a spark or flame.

Basic—Derivative of a compound that is more alkaline than other compounds of the same name, for example, lead carbonate, basic, basic salt.

Boiling Point—The temperature of a liquid at which its vapor pressure is equal to or very slightly greater than the atmospheric pressure of the environment.

Bonding—Coupling of two or more containers for the purpose of eliminating a potential charge difference between the containers during transport of flammable/combustible liquids.

Caustic—Any strongly alkaline material that has a corrosive or irritating effect on living tissue. When unqualified, this term usually refers to caustic soda (NAOH).

Combustible—Solids that are difficult to ignite and that burn relatively slowly, and liquids that have a flashpoint greater than 100°F (37.7°C).

Compatible—Materials that are able to exist in close and permanent association indefinitely.

Corrosion—The electrochemical degradation of metals or alloys due to reaction with their environment, that is accelerated by the presence of acids or bases.

Cryogenic—Liquefied gases at temperatures below -200°F.

Density—Weight per unit volume expressed as grams/cubic centimeter for solids and liquids and usually as grams/liter for gases.

Derivative—A substance that can be made from another substance in one or more steps.

Dilution—To diminish the strength, flavor, or brilliance by admixture.

Endothermic—A process or change that takes place with absorption of heat and requires high temperature for initiation and maintenance.

Exothermic—A process or chemical reaction that is accompanied by evolution (generation) of heat, for example, combustion reactions.

Fire Triangle—The idea that a fire cannot exist without all three critical components including heat, oxygen, and fuel. This does not include the idea of the chain reaction (Fire Tetrahedron).

Flammability Range—The range of vapor to air concentrations that will burn if an ignition source is present. This range has an *upper explosive limit* (UEL), or richest vapor-to-air mixture and a *lower explosive limit* (LEL) or leanest vapor-to-air mixture.

Flammable Material—Any solid, liquid, vapor, or gas that will ignite easily and burn rapidly. Flammable solids are of several types: (1) dusts or fine powders; (2) those that ignite spontaneously at low temperatures; (3) those in which internal heat is built up by microbial or other degradation activity; (4) films, fibers, and fabrics of low-ignition point materials.

Flash Point—(1) The temperature at which a liquid gives off enough vapor to form an ignitable mixture at the air liquid interface. (2) The minimum temperature at which a liquid gives off vapor in sufficient concentration to form an ignitable mixture with air near the surface of the liquid (see Title 49 CFR 173.115(d)).

Flashback—A recession of flame into an unwanted location.

Fumes—Particulate, smoke-like emanation from the surface of heated metals.

Grounding—Creating a large conducting body (as the Earth) used as a common return for an electric circuit and as an arbitrary zero of potential.

Insulator—Any substance or mixture that has an extremely low dielectric constant, low thermal conductivity, or both.

Lower Explosive Limit (LEL)—The leanest vapor-to-air concentration that will burn.

Lower Flammable Limit (LFL)—The lowest concentration of the material in air that will support combustion.

Necrosis—The death of living tissue (how quickly a substance will eat a hole in your hand).

Organic—Of, relating to, or containing carbon compounds.

Oxidizer—Any compound that spontaneously evolves oxygen either at room temperature or under slight heating.

PCB(s)—Polychlorinated biphenyl. A series of hazardous compounds used for a number of industrial purposes, which are now found throughout the natural environment. PCBs are toxic to some marine life at concentrations of a few parts per billion (ppb) and are known to cause skin diseases, digestive disturbances, and even death in humans at higher concentrations. PCBs are persistent in the environment and do not easily decompose, and biomagnify up the food chain.

PCP—(1) Abbreviation for pentachlorophenol, a wood preservative used on military ammunition boxes and telephone poles.

(2) 1-(1-Phenylcyclohexyl) piperidine or angel dust or HOG, an analgesic and anesthetic that may produce serious psychological disturbances.

pH—A numerical designation of relative acidity and alkalinity. A pH of 7.0 indicates precise neutrality; higher values indicate increasing alkalinity, and lower values indicate increasing acidity.

Physiology—The physical and chemical working of living things.

Poison—A substance that causes injury, illness, or death to living tissue by chemical activity.

Polymerize—A chemical reaction, usually carried out with a catalyst, heat, or light, and often under high pressure, in which a large number of relatively simple molecules combine to form a chain-like macromolecule.

Potential—Ability for something to occur.

ppb—Parts per billion (ppb); parts (by volume) of a gas or vapor, usually used to express measurements of extremely low concentrations of unusually toxic gases or vapors. It is also used to indicate the concentration of a particular substance in a liquid or solid.

ppm—Parts per million (ppm); refers to the concentration of a gas or vapor in air, also used to indicate the concentration of a particular substance in a liquid or solid. This unit is frequently written as milligram/liter (mg/L) for liquids and milligram/cubic meter (mg/m^3) for gases.

Pyrophoric—Any liquid or solid that will ignite spontaneously in air at about 130°F.

Radioactive—A material that emits alpha, beta, or gamma rays.

Reactivity—The ability of a material to react with ambient air or water.

Solubility—The ability or tendency of one substance to blend uniformly with another.

Specific Gravity—The ratio of the density of a substance to the density of a reference substance; it is an abstract number that is unrelated to any unit. The specific gravity of liquids and solids is the ratio of their density to that of water at 4°C, taken as 1.0.

Static Electricity—Charges created by the conductive movements of unlike materials.

Threshold Limit Value (TLV)—Threshold limit value. Registered trademark of the ACGIH. ACGIH presents the most recent TLVs for commonly used industrial chemical compounds.

Threshold Limit Value Ceiling—The concentration that should not be exceeded during any part of the working exposure.

Threshold Limit Value–Short-Term Exposure Limit (TLV-STEL, 15 minutes)—The concentration to which workers can be exposed continuously for a short period of time without suffering from (1) irritation; (2) chronic or irreversible tissue damage; or (3) narcosis of sufficient degree to increase the likelihood of accidental injury, impair self-rescue, or materially reduce work efficiency, provided that the daily TLV-TWA is not exceeded. This is not a separate independent exposure limit but, rather, supplements the time recognized acute effects from a substance whose toxic effects are primarily of a chronic nature. STELs are recommended only where toxic effects have been reported from high short-term exposures in either humans or animals.

Threshold Limit Value Time Weighted Average (TLV-TWA)—The time-weighted average concentration for a normal 8-hour workday and 40-hour workweek to which nearly all workers may be repeatedly exposed, day after day, without adverse effect.

Toxic—Any material that can cause harm by either ingestion, inhalation, or absorption via the skin. Classes of toxics:

a. Irritants—In very high concentrations could kill.
b. Asphyxiant—Gases that interfere with oxidation processes in the body.
c. Respiratory paralyzers—These short-circuit the respiratory nervous system.

Toxicity—Ability of a chemical, molecule, or compound to produce damage once it reaches a susceptible site in or near the body.

1. Acute toxicity results from a severe case of poisoning due to a single dose of exposure to a chemical.
2. Chronic toxicity is caused by repeated small doses over a considerable period (greater than 6 months), resulting in accumulation of the chemical in the body, or if its effects are additive, bringing about illness or sometimes death.
3. Dermal toxicity is a measure of the amount of a poison that can be absorbed through the skin of an animal or produce local toxicity. Measurements of toxicity are usually compared at the 50% level (see LD50).
4. Toxicity (Human). The AAPCO has adopted the following regulatory principles relating to the determination of highly toxic materials:
   a. Highly Toxic. An economic poison that falls within any of the following categories when tested on laboratory animals (mice, rats, and rabbits) and is highly toxic to humans within the meaning of these principles:

(1) "Oral Toxicity: Those materials which produce death in half or more than half the animals of any species at a dosage of 50 milligrams at a single dose, or less, per kilogram of body weight when administered orally to 10 or more such animals of each species."

(2) "Toxicity on inhalation: Those materials which produce death in half or more than half of the animals of any species at a dosage of 200 parts or less by volume of the gas or vapor per million parts by volume of air when administered by continuous inhalation for one hour or less to 10 or more animals of each special, provided such concentration is likely to be encountered by man when the economic poison is used in any reasonably foreseeable manner."

(3) "Toxicity by skin absorption: Those materials which produce death in half or more than half of the animals (rabbits only) tested a dosage of 200 milligrams or less per kilogram of body weight when administered by continuous contact with the bare skin for 24 hours or less to 10 or more animals. Provided, however, that an enforcement official may exempt any economic poison which meets the above standard but which is not in fact highly toxic to man, from these principles with respect to economic poisons highly toxic to man, and may after hearing designate as highly toxic to man any economic poison which experience has shown to be so in fact."

b. Toxicity Categories. EPA has published regulations for use of human hazard signal words on pesticide labels. The following signal words are assigned by levels of toxicity:

(1) Toxicity Category I. All pesticide products meeting the criteria of Toxicity Category I shall bear on the front panel the signal word "Danger." In addition, if the product has been assigned to Toxicity Category I on the basis of its oral, inhalation, or dermal toxicity (as distinct from skin and eye local effects), the word "Poison" shall appear in red on a background of distinctly contrasting color, and the skull and crossbones shall appear in immediate proximity to the word "Poison."

(2) Toxicity Category II. All pesticide products meeting the criteria of Toxicity Category II shall bear on the front panel the signal word "Warning."

(3) Toxicity Category III. All pesticide products meeting the criteria of Toxicity Category on the III shall bear on the front panel the signal word "Caution."

(4) Toxicity Category IV. All pesticide products meeting the criteria of Toxicity Category IV shall bear on the front panel the signal word "Caution."

(5) Use of signal words—Use of signal word(s) associated with a higher toxicity category is not permitted except when the agency determines that such labeling is necessary to prohibit unacceptable adverse effects.

Upper Explosive Limit (UEL)—(1) The richest vapor-to-air concentration that will burn. (2) The highest concentration of the material in air that can be detonated.

Upper Flammable Limit (UFL)—The highest concentration of the material in air that will support combustion.

Vapor Density—The ratio of the density of a gas relative to that of a reference gas. Ambient air is assigned a value of 1.0 (unitless).

Vapor Pressure—The pressure characteristic at any given temperature of a vapor in equilibrium with its liquid or solid form.

Volatility—Relative rate of evaporation of materials to assume the vapor state.

Current placards for Hazards Class 1 materials; note the associated labels will be the same.

Current placards for Hazards Class 2 and 3 materials; note the associated labels will be the same.

Current placards for Hazards Classes 4, 5, and 6 materials; note the associated labels will be the same.

Current placards for Hazards Classes 7, 8, and 9 materials; note the associated labels will be the same.

1.4.3 49 CFR Critical Definitions

This is the last set of definitions. The first set of definitions was the hazard classes; they are unique and extensive so they were presented separately. 49 CFR also defines many of the terms and a number of them can be anywhere from slightly to significantly different from the normally assumed definition for the term. Therefore this section adds some of the most critical and/or most atypical definitions from the 49 CFR. From a transportation standpoint these definitions would be essentially identical across the international and other state transportation regulations.

Bulk packaging means a packaging, other than a vessel or a barge, including a transport vehicle or freight container, in which hazardous materials are loaded with no intermediate form of containment and which has:

(1) A maximum capacity greater than 450 L (119 gallons) as a receptacle for a liquid;
(2) A maximum net mass greater than 400 kg (882 pounds) and a maximum capacity greater than 450 L (119 gallons) as a receptacle for a solid; or
(3) A water capacity greater than 454 kg (1000 pounds) as a receptacle for a gas as defined in § 173.115 of this subchapter.

Combination packaging means a combination of packaging, for transport purposes, consisting of one or more inner packagings secured in a non-bulk outer packaging. It does not include a composite packaging.

Commerce means trade or transportation in the jurisdiction of the United States within a single state; between a place in a state and a place outside of the state; that affects trade or transportation between a place in a state and place outside of the state; or on a United States-registered aircraft.

Consumer commodity means a material that is packaged and distributed in a form intended or suitable for sale through retail sales agencies or instrumentalities for consumption by individuals for purposes of personal care or household use. This term also includes drugs and medicines.

Domestic transportation means transportation between places within the United States other than through a foreign country.

Equivalent lithium content means, for a lithium-ion cell, the product of the rated capacity, in ampere-hours, of a lithium-ion cell times 0.3, with the result expressed in grams. The equivalent lithium content of a battery equals the sum of the grams of equivalent lithium content contained in the component cells of the battery.

Fuel cell means an electrochemical device that converts the energy of the chemical reaction between a fuel, such as hydrogen or hydrogen rich gases, alcohols, or hydrocarbons, and an oxidant, such as air or oxygen, to direct current (d.c.) power, heat, and other reaction products.

Fuel cell cartridge or *fuel cartridge* means an article that stores fuel for discharge into the fuel cell through a valve(s) that controls the discharge of fuel into the fuel cell.

Fuel cell system means a fuel cell with an installed fuel cell cartridge together with wiring, valves, and other attachments that connect the fuel cell or cartridge to the device it powers. The fuel cell or cartridge may be so constructed that it forms an integral part of the device or may be removed and connected manually to the device.

Fuel tank means a tank other than a cargo tank, used to transport flammable or combustible liquid, or compressed gas for the purpose of supplying fuel for propulsion of the transport vehicle to which it is attached, or for the operation of other equipment on the transport vehicle.

Gross weight or *gross mass* means the weight of a packaging plus the weight of its contents.

Hazardous material means a substance or material that the Secretary of Transportation has determined is capable of posing an unreasonable risk to health, safety, and property when transported in commerce, and has designated as hazardous under section 5103 of Federal hazardous materials transportation law (49 U.S.C. 5103). The term includes hazardous substances, hazardous wastes, marine pollutants, elevated temperature materials, materials designated as hazardous in the Hazardous Materials Table (see 49 CFR 172.101), and materials that meet the defining criteria for hazard classes and divisions in part 173 of subchapter C of this chapter.

Hazardous substance for the purposes of this subchapter, means a material, including its mixtures and solutions, that—

(1) Is listed in the appendix A to § 172.101 of this subchapter;
(2) Is in a quantity, in one package, which equals or exceeds the reportable quantity (RQ) listed in the appendix A to § 172.101 of this subchapter; and
(3) When in a mixture or solution—
 (i) For radionuclides, conforms to paragraph 7 of the appendix A to § 172.101.

	Concentration by weight	
RQ pounds (kilograms)	*Percent*	*PPM*
5000 (2270)	10	100,000
1000 (454)	2	20,000
100 (45.4)	0.2	2,000
10 (4.54)	0.02	200
1 (0.454)	0.002	20

 (ii) For other than radionuclides, is in a concentration by weight which equals or exceeds the concentration corresponding to the RQ of the material, as shown in table.

The term does not include petroleum, including crude oil or any fraction thereof which is not otherwise specifically listed or designated as a hazardous substance in appendix A to § 172.101 of this subchapter, and the term does not include natural gas, natural gas liquids, liquefied natural gas, or synthetic gas usable for fuel (or mixtures of natural gas and such synthetic gas).

Hazardous waste, for the purposes of this chapter, means any material that is subject to the Hazardous Waste Manifest Requirements of the U.S. Environmental Protection Agency specified in 40 CFR part 262.

HAZMAT means a hazardous material.

HAZMAT employee means:

(1) A person who is:
 (i) Employed on a full-time, part time, or temporary basis by a HAZMAT employer and who in the course of such full time, part time, or temporary employment directly affects hazardous materials transportation safety;

(ii) Self-employed (including an owner-operator of a motor vehicle, vessel, or aircraft) transporting hazardous materials in commerce who in the course of such self-employment directly affects hazardous materials transportation safety;
(iii) A railroad signalman; or
(iv) A railroad maintenance-of-way employee.

(2) This term includes an individual, employed on a full time, part time, or temporary basis by a HAZMAT employer, or who is self-employed, who during the course of employment:
(i) Loads, unloads, or handles hazardous materials;
(ii) Designs, manufactures, fabricates, inspects, marks, maintains, reconditions, repairs, or tests a package, container, or packaging component that is represented, marked, certified, or sold as qualified for use in transporting hazardous material in commerce;
(iii) Prepares hazardous materials for transportation;
(iv) Is responsible for safety of transporting hazardous materials;
(v) Operates a vehicle used to transport hazardous materials.

HAZMAT employer means:

(1) A person who employs or uses at least one HAZMAT employee on a full-time, part time, or temporary basis; and who:
(i) Transports hazardous materials in commerce;
(ii) Causes hazardous materials to be transported in commerce; or
(iii) Designs, manufactures, fabricates, inspects, marks, maintains, reconditions, repairs, or tests a package, container, or packaging component that is represented, marked, certified, or sold by that person as qualified for use in transporting hazardous materials in commerce;

(2) A person who is self-employed (including an owner-operator of a motor vehicle, vessel, or aircraft) transporting materials in commerce; and who:
(i) Transports hazardous materials in commerce;
(ii) Causes hazardous materials to be transported in commerce; or
(iii) Designs, manufactures, fabricates, inspects, marks, maintains, reconditions, repairs, or tests a package, container, or packaging component that is represented, marked, certified, or sold by that person as qualified for use in transporting hazardous materials in commerce; or

(3) A department, agency, or instrumentality of the United States Government, or an authority of a State, political subdivision of a State, or an Indian tribe; and who:
(i) Transports hazardous materials in commerce;
(ii) Causes hazardous materials to be transported in commerce; or
(iii) Designs, manufactures, fabricates, inspects, marks, maintains, reconditions, repairs, or tests a package, container, or packaging component that is represented, marked, certified, or sold by that person as qualified for use in transporting hazardous materials in commerce.

Household waste means any solid waste (including garbage, trash, and sanitary waste from septic tanks) derived from households (including single and multiple residences, hotels and motels, bunkhouses, ranger stations, crew quarters, campgrounds, picnic grounds, and day-use recreation areas). This term is not applicable to consolidated shipments of household hazardous materials transported from collection centers. A collection center is a central location where household waste is collected.

Incorporated by reference or *IBR* means a publication or a portion of a publication that is made a part of the regulations of this subchapter. See § 171.7.

Inner packaging means a packaging for which an outer packaging is required for transport. It does not include the inner receptacle of a composite packaging.

Inner receptacle means a receptacle which requires an outer packaging in order to perform its containment function. The inner receptacle may be an inner packaging of a combination packaging or the inner receptacle of a composite packaging.

Intermediate bulk container or *IBC* means a rigid or flexible portable packaging, other than a cylinder or portable tank, which is designed for mechanical handling. Standards for IBCs manufactured in the United States are set forth in subparts N and O of part 178 of this subchapter.

Intermediate packaging means a packaging which encloses an inner packaging or article and is itself enclosed in an outer packaging.

Intermodal container means a freight container designed and constructed to permit it to be used interchangeably in two or more modes of transport.

Intermodal portable tank or *IM portable tank* means a specific class of portable tanks designed primarily for international intermodal use.

Jerrican means a metal or plastic packaging of rectangular or polygonal cross section.

Large packaging means a packaging that—

(1) Consists of an outer packaging which contains articles or inner packagings;
(2) Is designated for mechanical handling;
(3) Exceeds 400 kg net mass or 450 liters (118.9 gallons) capacity;
(4) Has a volume of not more than 3 m^3 (see § 178.801(i) of this subchapter); and
(5) Conforms to the requirements for the construction, testing and marking of large packagings as specified in the UN Recommendations, Chapter 6.6 (incorporated by reference; see § 171.7).

Limited quantity, when specified as such in a section applicable to a particular material, means the maximum amount of a hazardous material for which there is a specific labeling or packaging exception.

Lithium content means the mass of lithium in the anode of a lithium metal or lithium alloy cell. The lithium content of a battery equals the sum of the grams of lithium content contained in the component cells of the battery. For a lithium-ion cell see the definition for "equivalent lithium content."

Loading incidental to movement means loading by carrier personnel or in the presence of carrier personnel of packaged or containerized hazardous material onto a transport vehicle, aircraft, or vessel for the purpose of transporting it, including the loading, blocking, and bracing a hazardous materials package in a freight container or transport vehicle, and segregating a hazardous materials package in a freight container or transport vehicle from incompatible cargo. For a bulk packaging, *loading incidental to movement* means filling the packaging with a hazardous material for the purpose of transporting it. *Loading incidental to movement* includes transloading.

Marking means a descriptive name, identification number, instructions, cautions, weight, specification, or UN marks, or combinations thereof, required by this subchapter on outer packagings of hazardous materials.

Material of trade means a hazardous material, other than a hazardous waste, that is carried on a motor vehicle—

(1) For the purpose of protecting the health and safety of the motor vehicle operator or passengers;

(2) For the purpose of supporting the operation or maintenance of a motor vehicle (including its auxiliary equipment); or
(3) By a private motor carrier (including vehicles operated by a rail carrier) in direct support of a principal business that is other than transportation by motor vehicle.

Non-bulk packaging means a packaging which has:

(1) A maximum capacity of 450 L (119 gallons) or less as a receptacle for a liquid;
(2) A maximum net mass of 400 kg (882 pounds) or less and a maximum capacity of 450 L (119 gallons) or less as a receptacle for a solid; or
(3) A water capacity of 454 kg (1000 pounds) or less as a receptacle for a gas as defined in § 173.115 of this subchapter.

Outage or *ullage* means the amount by which a packaging falls short of being liquid full, usually expressed in percent by volume.

Outer packaging means the outermost enclosure of a composite or combination packaging together with any absorbent materials, cushioning and any other components necessary to contain and protect inner receptacles or inner packagings.

Overpack, except as provided in subpart K of part 178 of this subchapter, means an enclosure that is used by a single consignor to provide protection or convenience in handling of a package or to consolidate two or more packages. *Overpack* does not include a transport vehicle, freight container, or aircraft unit load device. Examples of overpacks are one or more packages:

(1) Placed or stacked onto a load board such as a pallet and secured by strapping, shrink wrapping, stretch wrapping, or other suitable means; or
(2) Placed in a protective outer packaging such as a box or crate.

Package or *Outside Package* means a packaging plus its contents. For radioactive materials, see § 173.403 of this subchapter.

Packaging means a receptacle and any other components or materials necessary for the receptacle to perform its containment function in conformance with the minimum packing requirements of this subchapter. For radioactive materials packaging, see § 173.403 of this subchapter.

Packing group means a grouping according to the degree of danger presented by hazardous materials. Packing Group I indicates great danger; Packing Group II, medium danger; Packing Group III, minor danger. See § 172.101(f) of this subchapter.

Person means an individual, corporation, company, association, firm, partnership, society, joint stock company; or a government, Indian tribe, or authority of a government or tribe offering a hazardous material for transportation in commerce or transporting a hazardous material to support a commercial enterprise. This term does not include the United States Postal Service or, for purposes of 49 U.S.C. 5123 and 5124, a Department, agency, or instrumentality of the government.

Pre-transportation function means a function specified in the HMR that is required to assure the safe transportation of a hazardous material in commerce, including—

(1) Determining the hazard class of a hazardous material.
(2) Selecting a hazardous materials packaging.
(3) Filling a hazardous materials packaging, including a bulk packaging.

(4) Securing a closure on a filled or partially filled hazardous materials package or container or on a package or container containing a residue of a hazardous material.
(5) Marking a package to indicate that it contains a hazardous material.
(6) Labeling a package to indicate that it contains a hazardous material.
(7) Preparing a shipping paper.
(8) Providing and maintaining emergency response information.
(9) Reviewing a shipping paper to verify compliance with the HMR or international equivalents.
(10) For each person importing a hazardous material into the United States, providing the shipper with timely and complete information as to the HMR requirements that will apply to the transportation of the material within the United States.
(11) Certifying that a hazardous material is in proper condition for transportation in conformance with the requirements of the HMR.
(12) Loading, blocking, and bracing a hazardous materials package in a freight container or transport vehicle.
(13) Segregating a hazardous materials package in a freight container or transport vehicle from incompatible cargo.
(14) Selecting, providing, or affixing placards for a freight container or transport vehicle to indicate that it contains a hazardous material.

Proper shipping name means the name of the hazardous material shown in Roman print (not italics) in § 172.101 of this subchapter.

Shipping paper means a shipping order, bill of lading, manifest, or other shipping document serving a similar purpose and prepared in accordance with subpart C of part 172 of this chapter.

Storage incidental to movement means storage of a transport vehicle, freight container, or package containing a hazardous material by any person between the time that a carrier takes physical possession of the hazardous material for the purpose of transporting it in commerce until the package containing the hazardous material is physically delivered to the destination indicated on a shipping document, package marking, or other medium, or, in the case of a private motor carrier, between the time that a motor vehicle driver takes physical possession of the hazardous material for the purpose of transporting it in commerce until the driver relinquishes possession of the package at its destination and is no longer responsible for performing functions subject to the HMR with respect to that particular package.

(1) *Storage incidental to movement* includes—
   (i) Storage at the destination shown on a shipping document, including storage at a transloading facility, provided the shipping documentation identifies the shipment as a through-shipment and identifies the final destination or destinations of the hazardous material; and
   (ii) Rail cars containing hazardous materials that are stored on track that does not meet the definition of "private track or siding" in § 171.8, even if those cars have been delivered to the destination shown on the shipping document.
(2) Storage incidental to movement does not include storage of a hazardous material at its final destination as shown on a shipping document.

Undeclared hazardous material means a hazardous material that is: (1) Subject to any of the hazard communication requirements in subparts C (Shipping Papers), D (Marking), E (Labeling), and F

(Placarding) of Part 172 of this subchapter, or an alternative marking requirement in Part 173 of this subchapter (such as §§ 173.4(a)(10) and 173.6(c)); and (2) offered for transportation in commerce without any visible indication to the person accepting the hazardous material for transportation that a hazardous material is present, on either an accompanying shipping document, or the outside of a transport vehicle, freight container, or package.

(49 CFR, 2011)

1.5 Conversions Numbers, Densities, and Math

One of the biggest changes forced upon the United States in 1990 was the requirement to convert to the metric system for hazardous material/dangerous goods. The very nature of hazardous materials is heavily dependent on basic math and numbers; therefore, a review of American quantities and conversion charts to metric quantities would be very helpful. That is what you will find in Tables 1.2 through 1.9. Since many statutory requirements rely heavily on understanding on hazardous waste material and percentages, a review of this material is in order.

Table 1.2 U.S. Solid and Liquid Measurements

U.S. solid measurements		*U.S. liquid measurements*	
Normal expression	*Decimal equivalent*	*Normal expression*	*Decimal equivalent*
Ton	2000.0	Gallon	1.0
Pound	1.0	1/2 Gal	0.5
1/2 lb	0.5	Quart	0.25
1/4 lb	0.25	Pint	0.125
Ounce	0.0625	Ounce	0.0078125

Table 1.3 US 49 CFR Conversions

Multiply	*By*	*To obtain*	*Multiply*	*By*	*To obtain*
grams	0.03527	ounces	liters	0.2643	US gallons
grams	0.002205	pounds	liters	2.113	US pints
kilograms	35.2736	ounces	gallons	8	pints
kilograms	2.2046	pounds	US gallons	3.7853	liters
ounces	28.3495	grams	US pints	0.473	liters
pounds	16	ounces			
pounds	453.59	grams			
pounds	0.45359	kilograms			

Table 1.4 Mathematical Prefixes

pico	1 trillionth	1/1000,000,000,000
micro	1 millionth	1/1,000,000
milli	1 thousandth	1/1000
centi	1 hundredth	1/100
deci	1 tenth	1/10
deca	1 ten	10
kilo	1 thousand	1000
mega	1 million	1,000,000
giga	1 billion	1,000,000,000

1.5.1 Mass Conversion Table

When the central value in any row of these mass conversion tables (Tables 1.5 to 1.9) is taken to be in pounds, its equivalent value in kilograms is shown on the left; when the central value is in kilograms, its equivalent in pounds is shown on the right.

Table 1.5 U.S. to SI Conversions

To convert	*To*	*Multiply by*
pound avoirdupois	kilogram	0.4536
ounce avoirdupois	gram	28.35
gallon US liquid	liter	3.785
US quart	liter	0.9464
pint	liter	0.4732
ounce, fluid US	milliliter	29.57
pound per sq inch	kilopascal	6.895
curie	tetrabecquerel	0.037
rem	sievert	0.1000

Table 1.6 SI to U.S. Conversions

To convert	*To*	*Multiply by*
gram	ounce avoirdupois	0.03527
kilogram	pound	2.205
liter	gallon U.S. liquid	0.2642
liter	quart U.S.	1.057
liter	pint U.S.	2.113
milliliter	ounce fluid U.S.	0.03381
kilopascal	pounds/sq inch	0.1450
sievert	rem	100.00
tetrabecquerel	curie	27.03

Table 1.7 Mass Conversion Table

kg	← → *lb* *kg*	*lb*	*kg*	← → *lb* *kg*	*lb*	*kg*	← → *lb* *kg*	*lb*
0.227	0.5	1.10	22.7	50	110	90.7	200	441
0.454	1	2.20	24.9	55	121	95.3	210	463
0.907	2	4.41	27.2	60	132	99.8	220	485
1.36	3	6.61	29.5	65	143	102	225	496
1.81	4	8.82	31.8	70	154	104	230	507
2.27	5	11.0	34.0	75	165	109	240	529
2.72	6	13.2	36.3	80	176	113	250	551
3.18	7	15.4	38.6	85	187	118	260	573
3.63	8	17.6	40.8	90	198	122	270	595
4.08	9	19.8	43.1	95	209	125	275	606
4.45	10	22.0	45.4	100	220	127	280	617
4.99	11	24.25	47.6	105	231	132	290	639
5.44	12	26.5	49.9	110	242.5	136	300	661
5.90	13	28.7	52.2	115	254	159	350	772
6.35	14	30.9	54.4	120	265	181	400	882
6.80	15	33.1	56.7	125	276	204	450	992
7.26	16	35.3	59.0	130	287	227	500	1102
7.71	17	37.5	61.2	135	298	247	545	1202
8.16	18	39.7	63.5	140	309	249	550	1213
8.62	19	41.9	65.8	145	320	272	600	1323
9.07	20	44.1	68.0	150	331	318	700	1543
11.3	25	55.1	72.6	160	353	363	800	1764
13.6	30	66.1	77.1	170	375	408	900	1984
15.9	35	77.2	79.4	175	386	454	1000	2205
18.1	40	88.2	81.6	180	397			
20.4	45	99.2	86.2	190	419			

1.5.2 Liquid Conversions

1.5.2.1 Pints to Liters

Using the central column, it is possible to convert in either direction; choosing the appropriate numbers from the central column and understanding how to then multiply and add those numbers, one should be able to reach any finite quantity by moving the decimal point appropriately based upon the quantities you are working with. In some cases it may be necessary to do no more than moving the decimal and adding two numbers.

Table 1.8 Pints to Liters Conversion Table

Liters	← *Pint* → *Liter*	*Pints*
0.2365	0.5	1.0565
0.473	1.0	2.113
0.7095	1.5	3.1695
0.946	2.0	4.226
1.1825	2.5	5.2825
1.419	3.0	6.339
1.6555	3.5	7.3955
1.892	4.0	8.452
2.1285	4.5	9.5085
2.365	5.0	10.565
2.6015	5.5	11.6215
2.838	6.0	12.678
3.0745	6.5	13.7345
3.311	7.0	14.791
3.5475	7.5	15.8475
3.784	8.0	16.904
4.0205	8.5	17.9605
4.257	9.0	19.017
4.4935	9.5	20.0735
4.3	10.0	21.13

Two simple examples

105 pints would be equivalent to 43 L + 2.365 L or a total of 45.365 L.
73.5 L would be equal to 147.91 pints + 4.226 pints + 3.1695 pints or 155.3055 pints.

Table 1.9 Gallons to Liters Conversion Table

Gallons to liters								
Liter	*Gal lit*	*Gallon*	*Liter*	*Gal lit*	*Gallon*	*Liter*	*Gal lit*	*Gallon*
1.893	0.5	0.132	60.565	16	4.229	155.197	41	10.836
3.785	1	0.264	64.350	17	4.493	158.983	42	11.101
5.678	1.5	0.396	68.135	18	4.757	162.768	43	11.365
7.571	2	0.529	70.921	19	5.022	166.553	44	11.629
9.463	2.5	0.661	75.706	20	5.286	170.339	45	11.894
11.356	3	0.793	79.491	21	5.550	174.124	46	12.158
13.248	3.5	0.925	83.276	22	5.815	184.700	48	12.686
15.141	4	1.057	87.062	23	6.079	189.265	50	13.215
17.034	4.5	1.189	90.847	24	6.343	208.192	55	14.536
18.926	5	1.322	94.633	25	6.608	227.118	60	15.858
20.819	5.5	1.454	98.418	26	6.872	246.045	65	17.180
22.712	6	1.586	102.203	27	7.136	264.971	70	18.501
24.604	6.5	1.718	105.988	28	7.400	283.898	75	19.823
26.497	7	1.850	109.773	29	7.665	302.824	80	21.144
28.389	7.5	1.982	113.559	30	7.929	321.751	85	22.466
30.282	8	2.114	117.344	31	8.193	340.677	90	23.787
32.175	8.5	2.247	121.129	32	8.458	359.604	95	25.109
34.068	9	2.379	124.915	33	8.722	378.53	100	26.430
35.960	9.5	2.511	128.700	34	8.986	567.795	150	39.645
37.853	10	2.643	132.486	35	9.251	757.06	200	52.86
41.638	11	2.907	136.271	36	9.515	1135.59	300	79.29
45.424	12	3.172	140.056	37	9.779	1514.12	400	105.72
49.209	13	3.436	143.841	38	10.043	1892.65	500	132.15
52.994	14	3.700	147.627	39	10.308	3785.3	1000	264.3
56.779	15	3.965	151.412	40	10.572			

A practical application, determining whether a particular package meets the definition for reportable quantity under 49 CFR.

A quote from DOT training materials: "To be a hazardous substance, the material must meet or exceed the pound(s) or kilogram(s) per package as well as the concentration by weight as listed in the table (171.8)."

The following formula will help you to work through real-world problems and focus on the process of determining Reportable Quantities in solutions and mixtures.

PPM divided by 1,000,000 equals the % of substance.

% of substance × total volume equals % of volume
% of volume × density of material equals total wt of material

EXAMPLE

You have a 55 gallon drum full of a solution of benzene containing 8000 PPM of benzene. The weight of the drum's contents is 400 pounds.

1. 8000 divided by 1000000 = 0.008%
2. 0.008 × 55 = 0.44 gallon
3. 0.44 × density (7.3) = 3.21 pounds of benzene in this pkg.

RQ per package, from 49 CFR § 171.101 Table appendix A, is 10 pounds; therefore, this is not a hazardous substance.

References

Blanchard, B. (1974). *Logistics engineering and management.* New York: Prentice Hall.
U.S. Department of Labor (OSHA). (2011). 29 CFR.
U.S. DOT. (2011). 49 CFR.
U.S. EPA. (2011). 40 CFR.
U.S. GSA. (1999). FED-STD-313D with Change 1.

Chapter 2

International Regulatory Framework and Standards

2.1 Introduction

When all is said and done, it is the international framework that is applied today within the United States and elsewhere. When one talks about global or international economies, it is important to recognize that the word most commonly used to identify or attribute unique requirements is "state" rather than country. Each European country is a *state* in global terminology, and the nations that make up United Nations are referred to individually as states. That will create some problems in this text as everyone needs to remember that state has one meaning under US or federal regulations and a completely different meaning under all other circumstances and in all other places.

In the United States, major changes were made to the key transportation regulations in 1991. That was a direct result of the rest of the world saying no to the United States; in essence, the international community said the United States could no longer use its own definitions and its arbitrary system of measurements if they wished to export product to the rest of the world. The original United States hazardous materials transportation regulation is known as the Hazardous Materials Transportation Act of 1974 or HMT. The much more critical act is the 1990 Hazardous Materials Transportation Uniform Safety Act (HMTUSA).

While the material in this volume focuses very heavily on American legislation and American regulations, the concept and purpose is to create a viable and applicable framework. The underlying concepts and the science behind much of that is applicable anywhere in the world. Much of the material that appears in the second half of the book would be applicable in any setting regardless of specific national or local regulation. The dangers associated with materials we determine to be dangerous goods or hazardous materials remain and are based upon science regardless of whether or not there is enabling or limiting legislation within a state.

The short sections that follow explore the foundation for all the regulations, place some of the materials and ideas in historical context, and demonstrate the universality or commonality of the inherent need to understand and therefore regulate or control the use and disposal of such materials regardless of regulation.

2.2 United Nations (UN) Orange Book

Humanity has recognized the inherent threats from the dangers posed by various materials since at least the Middle Ages. Paracelsus, the Swiss-born Renaissance physician, best captured the key concept that drives environmental health safety and dangerous goods regulations today. It is he who said in 1543: "All things are poisonous, for there is nothing without poisonous qualities. It is only the dose that makes a thing poison" (Griffin, 2009). In 1661, John Evelyn published one of the more important texts, *Fumifugium*, which describes the impact of the industrialization that initially was known as the Industrial Revolution on the environment. In this work, he suggested it was necessary to outlaw high sulfur coal and move industrial processes such as iron foundries away from cities and into well-ventilated areas. He also was one of the earliest proponents of appropriate zoning (Griffin, 2009).

With the development of more industrial processes and the increased availability and desirability of mass-produced goods—which in turn engenders the use of more industrial processes and manufacturing facilities—many others have written texts that address the scientific and social issues of hazardous materials and air, water, and ground pollution at multiple levels. Various professional organizations have worked to create directions, guidance, and regulations, which in turn create a legally binding set of rules for the preparation, movement, and storage of hazardous materials. It was not until the mid-20th century that the transportation of hazardous materials became a serious enough problem that the global community recognized the need for more effective and more comprehensive regulation.

In North America alone, there have been three very significant transportation-related events in the first half of the 20th century:

1. The Halifax Explosion, the world's largest manmade explosion before Hiroshima, occurred when a vessel lost steerageway in the harbor and collided with a munitions carrier on December 6, 1917.
2. The Port Chicago explosion in July 1944 was another example of the dangers related to the movement of hazardous materials as well as being one of the low points in race relationships and professionalism in the U.S. Navy.
3. The Houston Ship Channel disaster, or, as it is more commonly known, the *Texas City Disaster*, was the deadliest industrial accident in U.S. history. The incident occurred on April 16, 1947.

There are dozens if not hundreds of other examples in North America alone and an equally large number of examples worldwide, but it was not until 1956 that the international community began a formal process to address these issues. The United States really did not fully understand the need for appropriate oversight and regulation until 1974 when the original Hazardous Materials Transportation Act was passed. As you can see, the basic legal and regulatory framework governing dangerous goods is a very recent phenomenon, and it is clearly undergoing major evolutionary changes as this is being written.

The original document that addresses this issue in a global manner is the work of United Nations Economic and Social Council's Committee of Experts on the Transport of Dangerous Goods titled "Recommendations on the Transport of Dangerous Goods." This work is universally referred to as the Orange Book. In 1996 this document was modified to include "model regulations"; in 1999 the Council officially expanded "the mandate of the committee to the global harmonization of the various systems specification labeling of chemicals which are applicable under various regulatory regimes, e.g.: transport; workplace safety; consumer protection; and environmental protection" (UN 2009). The cumulative effect of the original mandate in the 1996 and 1999 changes provides the basis for the discussion that appears at Section 3.2 in Chapter 3 of the text.

From both a legal and practical standpoint, this is the foundational document for materials that represent hazards to the environment and to living organisms. At the same time, it is both interesting and important to note that the United Nations does not have the legal authority to enforce such recommendations and they act, in essence, as a producer of consensus standards. It is only through the activities of either international or state organizations that enabling laws are drawn and regulations created. It is the Orange Book that forms the basis for the two statutory and one consensus standard that in turn govern international air and maritime transportation of dangerous goods, the International Civil Aviation Organization Technical Instructions (ICAO TI), The International Air Transport Association Dangerous Goods Reregulation (IATA DGR), and the International Maritime Organization Dangerous Goods Code (IMDGC). Overall management of these materials, conceptually it is very straightforward: four-digit numbers are assigned, and identifiers called "proper shipping name" are assigned. The universe of numbers runs from 0000 to roughly 3500. Not every number is used, and each number may have a series of different proper shipping names just as an individual proper shipping name may have more than one number. (Do not worry about that, just be aware of it; it should not have any impact on 98% of the individuals using this volume.) Another sidelight 95% to 96% of the readers do not have to worry about: there is an additional limited set of numbers used for air transportation. Called 10 numbers, or UN numbers as they are referred to universally, start with capital UN followed by four digits; in the past they used to be and a numbers which were unique to North America, but Canada and the United States eliminated all such numbers in the last 12 to 13 years. There are 150 ID numbers, which are always in the 8000 series and are preceded by the letters ID instead of UN. An ID 8000 series number represents a material that has not been through a full classification process and has not yet been assigned a UN number. The only place this is particularly critical is in air movement, and that is why we use and apply ID 8000 numbers to air regulations.

Let's get back to the basic concept of UN number and proper shipping name or PSN. There are no absolutes when it comes to numbering the original system-assigned numbers by class, and to a large extent that still holds true. Hazard class I, consisting of divisions 1.1 through 1.6, uses numbers from 0001 to 0509 at the current time. Note that there are always four digits of the number. The original basic numbering was as follows: Hazard class 2 comprises compressed gases and is divided into three divisions; it uses numbers from 1001 to 1087. Hazard class 3 comprises flammable liquids; the numbering starts with 1088 and runs to 1308. Hazard class 4 is divided into three divisions; it starts at 1309 and runs to 1437. Hazard class 5, consisting of two divisions, runs from 1438 to 1517. Hazard class 6 also consists of two divisions; it starts at 1541 and runs through 1713. There are some hazard materials in the range from 1714 to 1723: class 8 starts at 1724 running up to 1840, and class 9 starts with 1841 and originally also included 1845. Numbers from 1846 onwards were assigned after the original numbering was done, so there is no way to characterize the numbers from 1846 through 3496. The numbers above 3000 do begin to capture man-made or manufactured items rather than simple materials and include internal combustion engines, all batteries, fuel cells, EEBDs, and many other items.

2.3 International Maritime Organization (IMO) and the International Maritime Dangerous Goods Code (IMDGC)

While it may appear counterintuitive upon first consideration, this is one of the more restrictive and comprehensive regulations published. The IMDGC is only one work in a much larger body of work that regulates many of the issues related to the maritime sector. The International Maritime Organization (IMO) regulations cover environmental and maritime operational areas as diverse as ballast water exchange, hull antifouling, smokestack emissions, low-sulfur diesel fuels, and many other areas typically considered part of point source manufacturing and processing operations in the past.

Product, primarily raw and agricultural product, but either category, would not be shipped via other modes or shipped in small quantities over shorter distances, but greater amounts and longer times represent significant hazards when moved over the world's oceans. A parallel can be drawn with grain storage silos. The grain per se does not represent a hazard, but in the grain storage silo the mixture of extremely small particles in the ambient air within a silo creates an explosive hazard. There are similar situations with many different types of products when they are moved in bulk via ship. Coal dust is our next example. Coal does not represent the real hazard in shipment, but the coal dust in below-deck compartments can build up to create an explosive mixture. Other examples we do not usually think of in terms of certifying hazardous materials include UN1327 Hay; UN1345 Rubber Scrap; UN1361 Carbon, animal or vegetable origin: UN1363 Copra; UN1365 Cotton Wet; UN1372 Fiber or fabric animal, vegetable or synthetic; and UN1374 Fishmeal.

Marine transportation presents many challenges beyond what one normally thinks of, and therefore the IMDGC must consider issues that are of lesser concern in other modes. As an example, the sea state and the weather combine to exert extreme forces in three different planes at the same time, so packaging and packages require additional attention and consideration that might be ignored for other modes of transportation.

The large and growing volume of raw and finished goods moving on today's modern ocean liners and the unique challenges of moving dangerous goods classified as 4.3, "Dangerous When Wet," require the IMDGC to address areas not covered by any other mode. Other reasons for a more restrictive approach include the inherent dangers related to fire at sea and the relative proximity of crew as well as passengers on a seaborne platform. Some restrictions relate to a separation of all three planes, while others have to do with stowage above or below deck, and closer to or further away from spaces reserved for humans and human activity.

Beyond that, the area of packaging, which embraces much more than the individual packaged unit, is much more important from a purely ecological standpoint. When one uses the term *packaging* in the global business environment, one must include blocking, bracing, dunnage and pallets. Today, the largest majority of those materials are still made of organic materials, more specifically wood and fiberboard, what many Americans call paper or corrugated packages. Both wood and fiberboard can become hosts of any number of invasive and non-native species in the arachnid and nematode families from simple cockroaches to various types of destructive beetles. Wood that has not been treated or has been improperly treated can also carry diseases that destroy all sorts of plant life including fruits, vegetables, and ornamental or shade trees. Therefore, it is important for global businesses to understand the inherent threat and therefore liabilities of moving product from raw material and finished and tested/inspected complex subassemblies between jurisdictions. American businesses generally are unaware of the many individual state restrictions on movements of goods that apply to commercial and private vehicles. At the time of this writing it is illegal (and could prove to be extremely destructive and economically devastating) to move raw, untreated wood, such as firewood, in the form of wooden pallets or in any other form, out of Worcester County, MA, into any other jurisdiction because of an invasive beetle infestation.

2.4 International Civil Aviation Organization (ICAO) Technical Instructions and the International Air Transport Association (IATA) Dangerous Goods Regulations

The real challenge here is to understand each of the regulations and their interrelationships. There are some issues that are unique to transportation that we will cover, one being the fact that passengers and cargo are in close proximity. This is very different from highway and rail.

To start with a global discussion, the physical regulation which is most critical to all is the IATA DGR. This is a relatively unique regulation because it is an industry-developed and adopted, or "consensus" standard. That means it does not have the power of law behind it. At various times, it may be slightly more restrictive than the ICAO TI, or, depending primarily on actual publication schedules, a more prescriptive change in that publication may make it temporarily more restrictive. The IATA DGR devotes an entire section to "exceptions or variations" that are unique to member states. The legally binding document that may be used in international courts is the ICAO TI of the industry-standard or "working" document is the IATA DGR. This is not a small issue because it is the IATA DGR that mandates a very specific form used to certify hazardous material shipments by air.

For those shipping in the United States, most people would think to go to the 49 CFR, and that publication certainly does address any unique US requirements. When all the publications are standardized and harmonized, that will be the case, but since each of these regulations is issued or updated on a different schedule, it is wise to have access to the current version of all three.

Some unique features and issues include the concept that some materials may not be transported on the same aircraft as passengers. Another unique issue relates to magnetized material. Due to legacy navigation instrumentation and tools, magnetized material may present a unique challenge to an aircraft as a system; therefore, there are additional restrictions on the shipment magnetized material by air. On the other hand, the air transportation regulations treat new materials that have not yet been classified but required movement under regulatory requirements to be shipped using a unique series of identified which are the "ID 8000" numbers mentioned earlier. Those numbers may be assigned to a given material for what some would consider an extended period of time. You must recognize that the formal process and the extensive testing as well as international rules that must be in place before a new UN number could be assigned takes time. It is not the unilateral decision of the producer and a single state. All protocols must be followed, and that takes time.

The air regulations also are much more specific about the use of absorbent materials and linings because of the inherent threats associated with the closed environment for pressurized aircraft. That pressure also raises unique issues for many materials that one seldom makes note of for surface transportation. Because of unique considerations, there are a number of labels unique to air transportation. The labels most commonly associated with air transportation and dangerous goods are "cryogenic liquid," "keep away from heat," "magnetized material," "cargo aircraft only," and the "lithium battery handling" label. At the management level wants one to know that this category exists not that the manager will be able to name each label or understand when to apply it, that is the job of the operational staff.

To the author, the organization and structure of the packing instructions are uniquely different from other modes and make the most sense for those who fully understand all the issues related to the management of hazardous materials. Two specific examples of the logic behind design and assignment of packing groups are: Packing groups use three digits. The first digit always represents a class. If you see a packing group such as 402, you can be sure that it is for a hazard class four material. The regulations also recognize limited quantities and arbitrarily placed a Y in front of any packing instruction that addresses the legal shipment of limited quantities. These are small things, but they add up when one needs to train a segment of their workforce that is subject to constant turnover: Shipping and receiving functions ordinarily are entry points into an organization rather than career positions within an organization.

2.5 ADR, RID, and ADN

The European marketplace, well before the creation of the European Union, agreed upon a broadly accepted group of regulations for dangerous goods moving within Europe. For roadways they are known as the ADR, for the rail system they are known as the RID, and for European inland waterways they are known as the ADN. We will briefly discuss the ADR as many of the distinguishing features that separated US regulations would be similar to those found in the RID and the ADN. The importance here is that in recent years there has been a growing movement to require that shipments originating outside of a jurisdiction or trade zone must meet certain unique requirements for onward transportation within that trade zone. While we did not mention it earlier, that is yet another reason why global companies operating outside of North America need a basic understanding of US regulations. It is a good time to remind everybody that while the regulations serve a very real safety and security purpose, one of the underlying tenets of the creation of the regulations was to ensure the free flow of commerce, but only in a manner that ensures sufficient safety to all those in the transportation system and to the population at large. The chagrin felt by many about the onerous requirements often masks the fact that most of them are really there to assure the safety of these people and their employees, not just the anonymous "public."

There is one thing that makes the ADR stand out from the UN Orange Book, international regulations, and US regulations.

The ADR includes a unique identification scheme. Chapter 5.3 of the ADR addresses placarding and marking of containers, MEGCs, MEMUs, tank containers, portable tanks, and vehicles. In that chapter, a new additional identification system is identified and explained. We are talking about what is called the "Orange Colored Plate Marking." This is a complex but hazard-class-based system in which there is yet another column in the tables and another set of identifiers that focuses on identifying class information as distinct from unique material characteristics identification. It is possible that at some point in time instead of the current UN followed by four digits, we will see the hazard identification numbers, as they are more properly known, coupled to the four digit numeric that currently make up the second half of each UN number.

2.6 Harmonization and Standardization

The original effort was started within the UN Council for Economic Development quite some time ago, but the real impetus came about in 1989 when the US Congress passed the Hazardous Materials Transportation Uniform Safety Act or HMTUSA. For the next 5 or 6 years, the focus was on getting the various transportation regulations organized in a similar manner so that they looked more alike, making it easier for the reader to move between them and identify differences; and, getting the drafters of the regulations to move toward more standard definitions and requirements for Dangerous Goods: thus, the terms *harmonization* and *standardization*. In 1996 the council moved to expand the concept toward a Globally Harmonized System of Classification and

Labeling of Chemicals, a much larger concept. In 2003 the first Globally Harmonized System of Classification and Labelling of Chemicals (GHS) was published. The reasons are easiest to see and understand by looking at just two hazard classes and the various systems for defining or applying them as they exist outside of this global classification system document. Table 2.1 focuses on criteria for defining and identifying toxic materials at different levels. Table 2.2 is a comparison of the different regulations and standards that apply to flammable liquids.

While there are clearly challenges globally between countries, the United States faces multiple challenges given that there are seven standards in use in the country, with no two completely compatible, for toxic classification; and there are three classification standards for flammable liquids with no two completely compatible. Table 2.2 demonstrates why so many HAZMAT experts and HAZMAT trainers in the United States are spreading incorrect information. The simple explanation is that most are coming out of a discipline that only addresses one, or at the most two, of the standards, and most do not have the HAZMAT transportation background to recognize the full international scope of these issues.

Table 2.1 Classification Regulations or Standards for Toxic Materials

	Acute oral toxicity LD_{50} (mg/kg)					
Organization/country/ regulation or standard	*High*		*Hazard*			*Low*
	0…………		*< 50…………*	*< 500…………*		*< 5000…………*
ANSI/US/A 129.1	< 50 Highly Toxic		> 50 < 500 Toxic	> 500 < 2000 Harmful		
OSHA/US/HCS	< 50 Highly Toxic		> 50 < 500 Toxic			
EPA/US/FIFRA	0 ≤ 50 Toxicity Category I		> 50 ≤ 500 Toxicity Category II	> 500 < 5000 Toxic Category III		> 5000 Toxicity Category IV
CPSC/US/FHSA	< 50 Highly Toxic		> 50 ≤ 500 Toxic			
GHS	≤ 5	> 5 ≤ 50	> 50 ≤ 300	> 300 ≤ 2000	> 2000 ≤ 5000	
DOT/US	< 5 Picking Group 1	> 5 < 50 Picking Group II	> 50 < 200 (solid) > 50 > 500 (liquid) Picking Group III			
NFPA/US	≤ 5 Hazard Category 4	> 5 ≤ 50 Hazard Category 3	> 50 ≤ 500 Hazard Category 2	> 500 ≤ 2000 Hazard Category 1	> 2000 Hazard Category 0	
NPCA/US/HMIS	≤ 1 Toxicity Rating 4	> 1 ≤ 50 Toxicity Rating 3	> 50 ≤ 500 Toxicity Rating 2	> 500 ≤ 5000 Toxicity Rating 1		> 5000 Toxicity Rating 0
EU	< 25 Very Toxic	> 25 > 200 Toxic	> 200 < 2000 Harmful			
WHMIS/Canada	≤ 50 Very Toxic WHMIS Class D, Division 1, Subdivision A		> 50 ≤ 500 Toxic WHMIS Class D, Division 1, Subdivision B			
Australia/NOHSC	< 25 Very Toxic	> 25 < 200 Toxic	> 200 < 2000 Harmful			
Mexico	<1 Extremely Toxic	>20 < 50 Highly Toxic	> 50 < 500 Moderately Toxic	> 500 < 5000 Mildly Toxic		
Malaysia	< 25 Very Toxic		200 to 500 Harmful			
Japan	< 30 Poisonous			300 to 3000 Powerful		
Korea	< 25 Very Toxic	> 50 < 200 Toxic	> 200 < 2000 Harmful			

Source: U.S. OSHA, *A Guide to the Globally Harmonized System of Classfication and Labelling of Chemicals* (GHS), 2011, Figure 1.2.

Note: The numerical values on the hazard index scale in the table are not to scale.

Table 2.2 Classification Regulations or Standards for Flammable Liquids

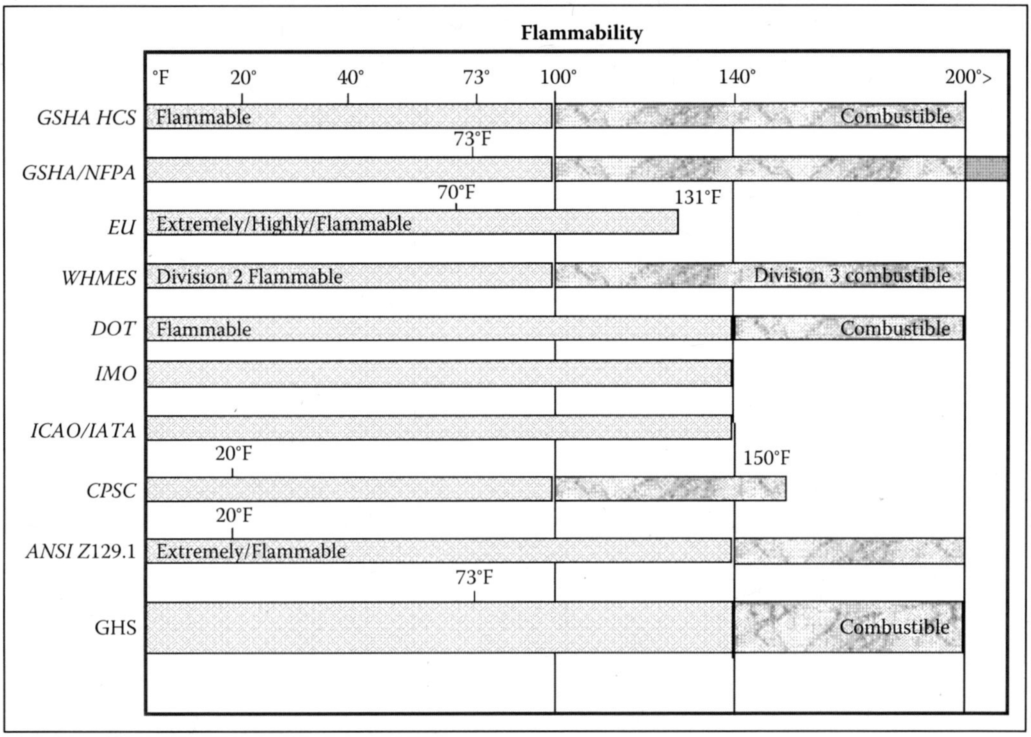

Flammability

	°F	20°	40°	73°	100°	140°	200°>
GSHA HCS	Flammable						Combustible
GSHA/NFPA				73°F			
EU	Extremely/Highly/Flammable			70°F		131°F	
WHMES	Division 2 Flammable					Division 3 combustible	
DOT	Flammable						Combustible
IMO							
ICAO/IATA							
CPSC		20°F				150°F	
*ANSI Z*129.1	Extremely/Flammable	20°F					
GHS				73°F			Combustible

Source: U.S. OSHA, *A Guide to the Globally Harmonized System of Classfication and Labelling of Chemicals* [GHS], 2011, Figure 1.3.

Note: The numerical values on the hazard index scale in the table are not to scale.

References

IATA. (2012). *Dangerous goods regulations.* Montreal, CA: IATA.

ICAO. (2010). *Technical instructions for the safe transportation of dangerous goods by air.* Montreal: ICAO.

IMO. (2010). *International maritime dangerous goods code.* London: IMO.

UN. (2009). *Recommendations on the transport of dangerous goods.* New York: UN.

U.S. OSHA. (2011). *A guide to the globally harmonized system of classification and labelling of chemicals* (GHS). Retrieved from http://www.osha.gov/dsq/hazcomm/qhs.html.

Chapter 3

U.S. Regulatory Framework and Standards

3.1 Overview of the Higher-Level Relationships in Government and Law

To be able to understand all the statutory requirements from multiple agencies and at multiple levels, one must be able to communicate and must have a fairly firm grasp of the political process within the United States. Each session of Congress enacts laws, many of which are captured and recognized as "acts"; one such act the author references regularly is the Energy Independence and Security Act (EISA) of 2007. Once an act has been passed by Congress, it must be written into U.S. code and, as a separate and parallel track, the code must be translated into a codified document that can be used by those outside of government. Most of the "laws" of the land are codified in a series of documents known as the CFRs.

The first problem here is that the U.S. Code and the Codes of Federal Regulations (CFRs) are not the same thing. The second issue is that the word "title" has more than one meaning when talking about federal regulations. There are 51 titles in the U.S. Code and there are 50 titled CFRs. Under the U.S. Code, Title 10 is Armed Forces, Title 11 is Bankruptcy, Title 29 is Labor, Title 32 is National Guard, Title 40 is Public Buildings Property and Works, Title 49 is Transportation, and Title 51 is National and Commercial Space Programs. When one looks at the "titles" for the CFRs, Title 10 is Energy, Title 11 is Federal Elections, Title 29 is Labor, Title 32 is National Defense, Title 40 is Protection of the Environment, Title 49 is Transportation, and there is no Title 51. At the same time, one must grasp that the public laws or "acts" passed by Congress are organized into titles so it is logical and standard practice to talk about something such as Title XI (note that we are now using Roman numerals) of the EISA of 2007; in this case we would be talking about "Energy, Transportation and Infrastructure."

The executive branch, through the designated cabinet-level agency or independent agency such as the EPA, has the responsibility to turn the laws or acts captured in the U.S. Code into, hopefully, usable "regulations." When the majority of the public talks about the regulations, they are talking about one of the 50 CFRs. If one is talking about normal enforcement activities through

FROM LAW TO REGULATION

TO GET FROM CONGRESS TO THE PUBLIC:

CABINET LEVEL DEPARTMENT GETS TASKING

OPERATING AGENCY GIVEN ASSIGNMENT

AGENCY WRITES THE CODE

Announcements in the federal register (ANPRM)

PUBLISHES IN FEDERAL REGISTER

Public Commentary

Final Rule in FEDERAL REGISTER

Annual publication of each volume of the CFR

GPO CONTRACTS FOR INITIAL PUBLICATION.

The 50 CFRs are divided into four groups and every CFR in a particular group gets issued/reissued at the beginning of that quarter in each fiscal year; 49 CFR is technically reissued at the beginning of the federal year, which is October 1. In practice the process is so complex and convoluted that is often the April/May time frame of the following calendar year before the actual CFR is signed off and printed.

Figure 3.1 From Congress to the public.

the operating agencies such as the EPA, OSHA, and the DOT modal administrations, one would typically be talking about the CFRs. On the other hand, if one is talking about formal proceedings instituted by the federal government in a court of law, then one is talking about lawyers and possibly a Department of Justice action, and there the US Code or USC would be referenced.

There is not even universal agreement as to how to write and properly show these documents when writing about them. The author is used to referring to a CFR by simply indicating the number followed by a space and then CFR: as an example, 49 CFR. Others feel it is necessary to add the word "title" in front of the number. In this text we will use a shortened version that will start with the actual number of the code, followed by CFR again: as an example, 40 CFR. Many professionals involved in various facets of environmental health, safety, security, and risk management regularly deal with all, or parts of, multiple regulations. The three most critical regulations are discussed in the text but that treatment is a very superficial treatment—just enough to allow the reader to develop sufficient knowledge so that they can move through the book and apply what is in the book to real-world work environments. Figure 3.1 is a diagrammatic representation of how congressional action translates eventually to the codified version of a CFR. It is the CFR that the business and private sectors at large use to determine how to conduct themselves and administer their organizations.

Many commercial sources offer these publications for sale in both hard copy and electronic versions. In some cases, purchasing from an outside or private source can be less expensive than purchasing from the government printing office (GPO).

3.2 FED-STD-313 and the Federal Acquisition Regulations (FAR)

The first step is to examine FED-STS-313 D. While this is a relatively short standard, careful examination and analysis of the standard reveals a lot. It is critical that one understands what is unstated and/or implied in this standard, as well as the language in the standard itself. Here are some key points that must be taken away from a careful reading and analysis of the standard:

a. The standard references include 11 external documents.
b. Those documents can be broken into three major groupings (although some documents may fall into multiple groups):
   Transportation regulations; six
   Federal environmental health and safety regulations; four
   Consensus standards; three
   One of the referenced publications, the USPS Publication 52, was redesignated years ago, and no longer provides guidance for the mailing of hazardous materials. Today, one would find that information in the domestic mail manual within DMM 600 Basic Standards for All Mailing Services, more specifically at 601, Mailability, in Section 10. That change is probably close to a decade old at this point in time, yet the federal standard has never been updated or corrected. One would need to know that such a change occurred and where to look today to find such information. At that point of discovery, a copy of the federal standard should be annotated to ensure its accuracy or it could be lost again. Many organizations and mailing addresses listed in the federal standard have probably changed since it was last updated, but this is of no real significance when discussing the topic as distinct from the unique agency point of contact.
c. The actual definition of hazardous materials, for the purpose of doing business with the federal government or as used by the federal government itself, takes a full page and encompasses a very large number of definitions that can be found in eight of the referenced documents. It should also be noted that some of these definitions contradict each other.
d. "Hazardous substance" has a unique definition, one that is distinct from the definition of "hazardous material."
e. "Hazardous waste" has a unique definition, one that is distinct from the definition of "hazardous material."
f. Table II in Appendix A of the standard, as shown, demonstrates how many different groups of products and materials fall within the realm of regulated materials. This table underlines how pervasive hazardous materials are in an industrial and developed economy (Table 3.1).
   A brief review of the groups of materials and the specific examples listed that require a material safety data sheet should clearly demonstrate that there is no facet of personal life or business that does not include hazardous materials. While many of these materials might appear to be unregulated within their normal setting, they are, in fact, regulated. That leads to another key concept of "regulated materials," which is a critical concept when talking about hazardous materials. That term will be explored in greater depth in the next section.

Table 3.1 Table of Federal Supply Groups and Hazardous Materials Table II of Appendix A in FED-STD-313D

FSG	*Title*	*Examples of HM*
12	Fire Control Equipment	Initiator propellants, cartridges power device
13	Ammunition and Explosives	Explosive devices, fire starter, flares
14	Guided Missile Components and Accessories	Cartridges power device, rockets, PCBs
15	Aircraft and Airframe Structural Components	Radioactive materials
16	Aircraft Components and Accessories	Items containing asbestos
22	Railway Equipment	Items containing asbestos
25	Vehicular Equipment Components	Items containing asbestos
26	Tires and Tubes	Items containing flammable or toxic compounds
34	Metalworking Machinery	Compressed gases, cleaners, acids, flux, and supplies containing or producing hazardous fumes
36	Miscellaneous Machinery	Flammable or toxic hazardous fumes
42	Fire Fighting, Rescue, and Safety Equipment	Extinguishing agents, repair and refill kits containing hazardous chemicals, items containing compressed gases or initiating charges
53	Hardware and Abrasives	Asbestos material, lead caulking, hazardous chemicals, items producing hazardous dust
54	Prefabricated Structures and Scaffolding	Repair kits containing hazardous chemicals
56	Construction and Building Materials	Cutback asphalt, deck and floor covering, sealing compounds, asbestos, formaldehyde, repair kits containing hazardous chemicals
58	Communication, Detection, and Coherent Radiation Equipment	Circuit cooler items containing ozone depleting substances, cleaners with hazardous chemicals
59	Electrical and Electronic Equipment Components	Items with PCBs, radioactive materials, flammable solvents, asbestos, or magnetic items
62	Lighting Fixtures and Lamps, Household and Quarters	Use items containing mercury or radioactive materials
63	Alarm Signal and Security Detection Systems	Items containing wet batteries or radioactive materials
65	Medical, Dental, and Veterinary Equipment and Supplies	Items containing hazardous chemicals, radioactive materials, mercury, asbestos, or flammable solvents

Table 3.1 *(Continued)* Table of Federal Supply Groups and Hazardous Materials Table II of Appendix A in FED-STD-313D

FSG	*Title*	*Examples of HM*
66	Instruments & Laboratory Equipment	Radioactive materials, flammable compounds, mercury, asbestos, compressed gases
67	Photographic Equipment	Radioactive compounds, solvents, thinners, and cements
75	Office Supplies & Devices	Solvents, thinners, cleaning fluids, flammable inks, varnishes, and chemicals which off-gas
84	Clothing, Individual Equipment, & Insignia	Maintenance kits containing flammable solvents
85	Personal Toiletry Articles	Pressurized containers with flammable or nonflammable propellants
87	Agricultural Supplies	Items containing herbicides and/or insecticides
93	Fabricated Materials	Items containing flammable solvents or toxic materials
96	Ores, Minerals, & Their Primary Products	Asbestos, mica, silica

g. A careful examination of FED-STD-313D and change 1 to the standard reveals a clear weakening of the standard itself. That change was executed by and promulgated by a minority group within the original body that drafted the original FED-STD-313D.

The development the promulgation of FED-STD-313D forms the basis and the rationale for the development of the rest of this work.

The concept of hazardous material and its definition as derived from the FAR, which in turn leads you to FED-STD-313D as the definition for hazardous material and is the basis for this work.

3.3 Transportation as the Basic Framework for the Text

There are a number of reasons why transportation is considered the basis for this text.

Many hazardous materials are recovered and used in extremely large quantities and/or are unsuitable for most uses in the form they are initially produced as recovered. All those materials must be moved multiple times. In all cases, starting from raw material to consumer or business user, hazardous materials often exist in a raw or unprocessed form and must first be delivered to a processing facility. They then need to go through the distribution chain and eventually show up at the user site. That tells us that many hazardous materials spend more time in transit than they do in production or use, and that without transportation there would be no hazardous materials for us to use. Most of these hazardous materials travel extremely long distances, across multiple continents in many cases, before they become the product that appears in a business or in the home. That alone makes an excellent case for using transportation as the basic framework.

A second reason, closely related to the last part of that first paragraph, is that a very large number of specific hazardous materials are produced in every country, which immediately tells us that they need to move within countries and over the surface within entire countries and between countries. That creates a need for an internationally acceptable standard. The reason is lost on many but in the end, as you will see repeatedly within this text, regulations and standards are put in place to first to ensure the free flow of commerce and, second, in regards to dangerous goods, with minimal, or theoretically no risk posed to the environment or humans and other living things as a result of the preparation or movement of those goods.

All the issues related to global harmonization and standardization grew out of the late 20th century work of the UN Council for Economic Development in the area of dangerous goods transportation and in 1996 that universally accepted work was expanded to include classification of all such materials in previously separate disciplines/areas.

Since the late 1980s, those properly and correctly trained on the preparation packaging and certification of dangerous goods for shipment have been exposed to and trained to the various (although now coalescing) international regulatory standards.

Figure 3.2 requires some discussion for full understanding. The concept can also be extended into additional areas relating to substances regulated strictly under environmental regulations.

This diagram is derived from the 49 CFR. In essence, the 49 CFR by its own definition of hazardous material is a stand-alone document. Within that definition 49 CFR recognizes the concept that anything considered hazardous substance under the 40 CFR *must* be treated as a hazardous material for purposes of a transportation. In this manner, the 49 accepts external definitions for materials that represent some form of hazard and therefore must be regulated. It is in the 40 CFR that the distinction is made about hazardous wastes. By definition, a hazardous waste is a hazardous substance; that does not mean every hazardous substance is hazardous waste, and that is what the diagram tries to capture. Many materials may qualify under all three

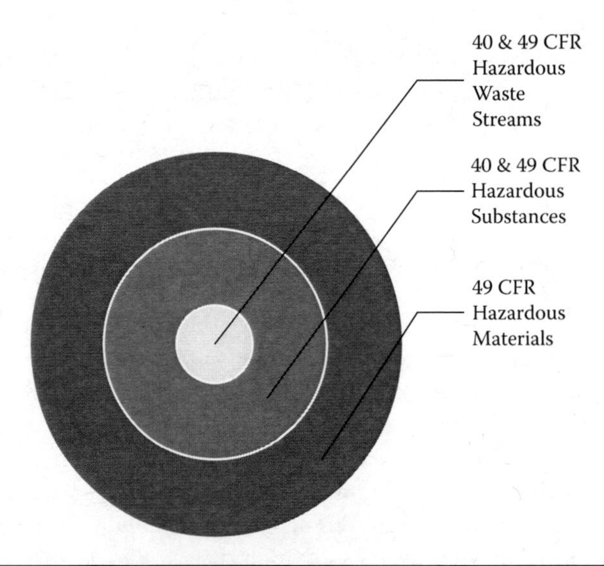

Figure 3.2 The unique relationship between key terms in the United States based upon federal regulations and the underlying UN recommendations.

definitions from the two regulations. The material may be a hazardous waste regardless of its former concentration or, in its purest state, it may be a hazardous substance while being a hazardous material at the same time. In some cases, the chemical or physical form and state of the material may change its definition or definitions. The state of a material becomes critical when we begin to talk about hazardous substances. Certain materials as observable solids may not be considered hazardous, yet if they exist in small enough sizes in certain physical states (the term most commonly used is *frangible*), then they become subject to inhalation and therefore become hazardous. While discussing this is going slightly afield of the main topic, the whole concept of anthrax and many other "weaponized" materials is based upon the danger of fragments of what is otherwise considered solid material.

This is yet another reason that transportation serves as such a logical basis for the treatment and discussion and management of all forms of materials that are hazardous regardless of the name used: dangerous goods, hazardous materials, hazardous substances, hazardous wastes. The transportation function. The supporting regulations and standards inherently and intrinsically embrace the material labeled and defined as hazardous by any of the regulation and, at the same time, addresses all of them in a similar manner for the purposes of transport.

3.4 49 CFR

3.4.1 § 171.7 Reference Material

(a) *Matter incorporated by reference*—(1) *General.* There is incorporated, by reference in parts 170–189 of this subchapter, matter referred to that is not specifically set forth. This matter is hereby made a part of the regulations in parts 170–189 of this subchapter. The matter subject to change is incorporated only as it is in effect on the date of issuance of the regulation referring to that matter. The material listed in paragraph (a)(3) has been approved for incorporation by reference by the Director of the Federal Register in accordance with 5 U.S.C 552(a) and 1 CFR part 51. Material is incorporated as it exists on the date of the approval and a notice of any change in the material will be published in the Federal Register. Matters referenced by footnote are included as part of the regulations of this subchapter.

(2) *Accessibility of materials.* All incorporated matter is available for inspection at:

(i) The Office of Hazardous Materials Safety, Office of Hazardous Materials Standards, East Building, PHH–10, 1200 New Jersey Avenue, SE., Washington, DC 20590–0001. For information on the availability of this material at PHH–10, call 1–800–467–4922, or go to: *http://www.phmsa.dot.gov*; and

(ii) The National Archives and Records Administration (NARA). For information on the availability of this material at NARA, call 202–741–6030, or go to: *http://www.archives.gov/federal_register/code_of_federal_regulations/ibr_locations.html*

(3) *Table of material incorporated by reference.* The following table sets forth material incorporated by reference. The first column lists the name and address of the organization from which the material is available and the name of the material. The second column lists the section(s) of this subchapter, other than § 171.7, in which the matter is referenced. The second column is presented for information only and may not be all inclusive.

Source and name of material	*49 CFR reference*
Air Transport Association of America, 1301 Pennsylvania Avenue, NW, Washington, DC 20004–1707	
The Aluminum Association, 420 Lexington Avenue, New York, NY 10017, telephone 301–645–0756, http://www.aluminum.org	
American National Standards Institute, Inc., 25 West 43rd Street, New York, NY 10036	
American Petroleum Institute, 1220 L Street, NW, Washington, DC 20005–4070	
American Pyrotechnics Association (APA), P.O. Box 30438, Bethesda, MD 20824, (301) 907–8181, http://www.americanpyro.com	
American Society of Mechanical Engineers, ASME International, 22 Law Drive, P.O. Box 2900, Fairfield, NJ 07007–2900, telephone 1–800–843–2763 or 1–973–882–1170, http://www.asme.org	
American Society for Testing and Materials, 100 Barr Harbor Drive, West Conshohocken, PA 19428, telephone (610) 832–9585, http://www.astm.org	
American Water Works Association, 1010 Vermont Avenue, NW, Suite 810, Washington, DC 20005	
American Welding Society, 550 NW Le Jeune Road, Miami, Florida 33126	
Association of American Railroads, American Railroads Building, 50 F Street, NW, Washington, DC 20001; telephone (877) 999–8824, http://www.aar.org/publications.com	
Chlorine Institute, Inc., 1300 Wilson Boulevard, Arlington, VA 22209	
Canadian General Standards Board, Place du Portage III, 6B1 11 Laurier Street, Gatineau, Quebec, Canada K1A 1G6	
Compressed Gas Association, Inc., 4221 Walney Road, 5th Floor, Chantilly, Virginia 20151, telephone (703) 788–2700, http://www.cganet.com	
Department of Defense (DOD), 2461 Eisenhower Avenue, Alexandria, VA 22331	
Department of Energy (USDOE), 100 Independence Avenue SW, Washington, DC 20545	

Source and name of material	*49 CFR reference*
General Services Administration, Specification Office, Room 6662, 7th and D Streets SW, Washington, DC 20407	
Institute of Makers of Explosives, 1120 19th Street NW, Suite 310, Washington, DC 20036–3605	
International Atomic Energy Agency (IAEA), P.O. Box 100, Wagramer Strasse 5, A–1400 Vienna, Austria	
International Civil Aviation Organization (ICAO), 999 University Street, Montréal, Quebec, Canada H3C 5H7, 1–514–954–8219, http://www.icao.int	
International Electrotechnical Commission (IEC) 3, rue de Varembé, P.O. Box 131, CH—1211, Geneva 20, Switzerland	
International Maritime Organization (IMO), 4 Albert Embankment, London, SE1 7SR, United Kingdom or New York Nautical Instrument & Service Corporation, 140 West Broadway, New York, NY 10013, +44 (0) 20 7735 7611, http://www.imo.org	
International Maritime Dangerous Goods Code (IMDG Code), 2008 Edition, Incorporating Amendment 34–08 (English Edition), Volumes 1 and 2	171.22; 171.23; 171.25; 172.101 Appendix B; 172.202; 172.401; 172.502; 172.602; 173.21; 173.56; 176.2; 176.5; 176.11; 176.27; 176.30; 176.84; 178.3; 178.274.
International Organization for Standardization, Case Postale 56, CH–1211, Geneva 20, Switzerland, http://www.iso.org	
National Board of Boiler and Pressure Vessel Inspectors, 1055 Crupper Avenue, Columbus, Ohio 43229	
National Fire Protection Association, Batterymarch Park, Quincy, MA 02269	
National Institute of Standards and Technology, Department of Commerce, 5285 Port Royal Road, Springfield, VA 22151	
Organization for Economic Cooperation and Development (OECD), OECD Publications and Information Center, 2001 L Street NW, Suite 700, Washington, DC 20036	
Transport Canada, TDG Canadian Government Publishing Center, Supply and Services, Ottawa, Ontario, Canada K1A 059, 416–973–1868, http://www.tc.gc.ca	

Source and name of material	*49 CFR reference*
Truck Trailer Manufacturers Association, 1020 Princess Street, Alexandria, Virginia 22314	
United Nations, Publications, 2 United Nations Plaza, Room DC2–853, New York, NY 10017, 1–212–963–8302, http://unp.un.org	
UN Recommendations on the Transport of Dangerous Goods, Fifteenth revised edition (2007). Volumes I and II	171.8; 171.12; 171.22; 171.23; 172.202; 172.401; 172.502; 173.22; 173.24; 173.24b; 173.40; 173.56; 173.192; 173.197; 173.302b; 173.304b; 178.75; 178.274; 178.801.
UN Recommendations on the Transport of Dangerous Goods, Manual of Tests and Criteria, Fourth revised edition (2003), and Addendum 2 (2004)	172.102; 173.21; 173.56; 173.57; 173.58; 173.115; 173.124; 173.125; 173.127; 173.128; 173.137; 173.185; Part 173, appendix H; 178.274.
United States Enrichment Corporation, Inc. (USEC)	

(b) *List of informational materials not requiring incorporation by reference.* The materials listed in this paragraph do not require approval for incorporation by reference and are included for informational purposes. These materials may be used as noted in those sections in which the material is referenced.

Source and name of material	*49 CFR reference*
American Biological Safety Association, 1202 Allanson Road, Mundelein, IL 60060	
Risk Group Classification for Infectious Agents, 1998	173.134
American Institute of Chemical Engineers (AIChE), 3 Park Avenue, New York, NY 10016–5991	
Process Safety Progress Journal, Vol. 21, No. 2	
Example of a Test Method for Venting Sizing: OPPSD/SPI Methodology	Note to § 173.225(h)(3)(vi)
American Society for Testing and Materials, 100 Barr Harbor Drive, West Conshohocken, PA 19428 Noncurrent ASTM Standards are available from: Engineering Societies Library, 354 East 47th Street, New York, NY 10017	
ASTM E 380–89 Standards for Metric Practice	171.10

Source and name of material	*49 CFR reference*
Association of American Railroads, American Railroads Building, 50 F Street NW, Washington, DC 20001	
AAR Catalog Nos. SE60CHT; SE60CC; SE60CHTE; SE60CE; SE60DC; SE60DE	179.14
AAR Catalog Nos. SE67CC; SE67CE; SE67BHT; SE67BC; SE67BHTE; SE67BE	179.14
AAR Catalog Nos. SE68BHT; SE68BC; SE68BHTE; SE68BE	179.14
AAR Catalog Nos. SE69AHTE; SE69AE	179.14
AAR Catalog Nos. SF70CHT; SF70CC; SF70CHTE; SF70CE	179.14
AAR Catalog Nos. SF73AC; SF73AE; SF73AHT; SF73AHTE	179.14
AAR Catalog Nos. SF79CHT; SF79CC; SF79CHTE; SF79CE	179.14
Bureau of Explosives (BOE), Hazardous Materials Systems, Association of American Railroads, American Railroads Building, 50 F Street NW, Washington, DC 20001	
Fetterley's Formula (The determination of the relief dimensions for safety valves on containers in which liquefied gas is charged and when the exterior surface of the container is exposed to a temperature of 1,200°F.)	173.315
Pamphlet 6, Illustrating Methods for Loading and Bracing Carload and Less-Than-Carload Shipments of Explosives and Other Dangerous Articles, 1962	174.55; 174.101; 174.112; 174.115; 174.290
Pamphlet 6A (includes appendix No. 1, October 1944, and appendix 2, December 1945), Illustrating Methods for Loading and Bracing Carload and Less-Than-Carload Shipments of Loaded Projectiles, Loaded Bombs, etc., 1943	174.101; 174.290
Pamphlet 6C, Illustrating Methods for Loading and Bracing Trailers and Less-Than-Trailer Shipments of Explosives and Other Dangerous Articles Via Trailer-on-Flatcar (TOFC) or Container-on-Flatcar (COFC), 1985	174.55; 174.63; 174.101; 174.112; 174.115
Emergency Handling of Hazardous Materials in Surface Transportation, 1989	171.7
Centers for Disease Control and Prevention 1600 Clifton Road, Atlanta, GA 30333	
Biosafety in Microbiological and Biomedical Laboratories, Fourth Edition, April 1999	173.134
Compressed Gas Association, Inc., 4221 Walney Road, 5th Floor, Chantilly, Virginia 20151	

Source and name of material	*49 CFR reference*
CGA C–1.1, Personnel Training and Certification Guidelines for Cylinder Requalification By the Volumetric Expansion Method, 2004, First Edition	180.209
National Institutes of Health Bethesda, Maryland 20892	
NIH Guidelines for Research Involving Recombinant DNA Molecules (NIH Guidelines), January 2001, Appendix B	173.134
Pantone Incorporated 590 Commerce Boulevard, Carlstadt, New Jersey 07072–3098	
Pantone® Formula guide coated/uncoated, Second Edition, 2004	172.407, 172.519
Society of Plastics Industries, Inc., Organic Peroxide Producers Safety Division, 1275 K Street NW, Suite 400, Washington, DC 20005	
Self Accelerating Decomposition Temperature Test, 1972	173.21
Truck Trailer Manufacturers Association, 1020 Princess Street, Alexandria, Virginia 22314; telephone: (703) 549–3010, http://www.ttmanet.org	
TTMA RP No. 96–01, TTMA RP No. 96–01, Structural Integrity of DOT 406, DOT 407, and DOT 412 Cylindrical Cargo Tanks, January 2001 Edition	178.345–3

3.5 40 CFR

The 40 CFR is the most ungainly of the regulations that all businesses which produce goods or services must deal with.

It is a huge regulation. In fact, it is really a compendium of regulatory requirements and processes that represent dozens of separate federal laws. For the purposes of this work only, the most critical laws, more commonly referred to as "acts," that Congress has passed and amended are very briefly outlined. The 40 CFR itself consists of 33 separate printed volumes which, in turn, are organized into seven separate chapters. As we have learned, all regulations are typically referred to in terms of the "part and section"; The 40 CFR contains 1799 distinct parts.

The major acts which we will touch upon that are included in this comprehensive regulation are the:

Clean Air Act (CAA)
Clean Water Act (CWA)
Comprehensive Environmental Response Compensation and Liability Act (CERCLA)

Emergency Planning and Community Right to Know Act (EPCRA) Also Known As Sara Title III
And, the Resource Conservation and Recovery Act (RCRA), which includes the critical Toxics Release Inventory (TRI) Program.

The salient points and key information from each of these acts are excerpted and summarized in the sections to follow.

3.5.1 Extracted and Edited Material From the EPA: "Plain English Guide to the Clean Air Act"

3.5.1.1 Understanding the Clean Air Act

3.5.1.1.1 Brief History of the Clean Air Act

In October 1948, a thick cloud of air pollution formed above the industrial town of Donora, Pennsylvania. The cloud, which lingered for five days, killed 20 people and caused sickness in 6,000 of the town's 14,000 people. In 1952, over 3,000 people died in what became known as London's "Killer Fog." The smog was so thick that buses could not run without guides walking ahead of them carrying lanterns.

Events like these alerted us to the dangers that air pollution poses to public health. Several U.S. federal and state laws were passed, including the original Clean Air Act of 1963. But there was no comprehensive federal response to address air pollution until Congress passed a much stronger Clean Air Act in 1970. That same year, Congress created the EPA and gave it the primary role in carrying out the law. Since 1970, EPA has been responsible for a variety of Clean Air Act programs to reduce air pollution nationwide.

In 1990, Congress dramatically revised and expanded the Clean Air Act, providing EPA even broader authority to implement and enforce regulations reducing air pollutant emissions. The 1990 Amendments also placed an increased emphasis on more cost-effective approaches to reduce air pollution.

3.5.1.1.2 Clean Air Act Roles and Responsibilities

States, tribes, and local governments do a lot of the work to meet the act's requirements. For example, representatives from these agencies work with companies to reduce air pollution. They also review and approve permit applications for industries or chemical processes.

3.5.1.1.3 EPA's Role

EPA sets limits on certain air pollutants, including limits on how much can be in the air anywhere in the United States. The Clean Air Act also gives EPA the authority to limit emissions of air pollutants coming from sources like chemical plants, utilities, and steel mills. Individual states or tribes may have stronger air pollution laws, but they may not have weaker pollution limits than those set by EPA.

EPA must approve state, tribal, and local agency plans for reducing air pollution. If a plan does not meet the necessary requirements, EPA can issue sanctions against the state and, if necessary, take over enforcing the Clean Air Act in that area.

EPA assists by providing research, expert studies, engineering designs, and funding to support clean air progress. Since 1970, Congress and the EPA have provided several billion dollars to the states, local agencies, and tribal nations to accomplish this.

3.5.1.1.4 State and Local Governments' Role

It makes sense for state and local air pollution agencies to take the lead in carrying out the Clean Air Act. They are able to develop solutions for pollution problems that require special understanding of local industries, geography, housing, and travel patterns, as well as other factors.

State, local, and tribal governments also monitor air quality, inspect facilities under their jurisdictions, and enforce Clean Air Act regulations.

States have to develop State Implementation Plans (SIPs) that outline how each state will control air pollution under the Clean Air Act. A SIP is a collection of the regulations, programs, and policies that a state will use to clean up polluted areas. The states must involve the public and industries through hearings and opportunities to comment on the development of each state plan.

3.5.1.1.5 Tribal Nations' Role

The 1990 revision of the Clean Air Act, recognizes that Indian tribes have the authority to implement air pollution control programs. Tribal Authority Rule gives tribes the ability to develop air quality management programs, write rules to reduce air pollution, and implement and enforce their rules in Indian country. While state and local agencies are responsible for all Clean Air Act requirements, tribes may develop and implement only those parts of the Clean Air Act that are appropriate for their lands.

3.5.1.2 Key Elements of the Clean Air Act

3.5.1.2.1 Cleaning Up Commonly Found Air Pollutants

Six common air pollutants (also known as "criteria pollutants") are found all over the United States. They are particle pollution (often referred to as particulate matter), ground-level ozone, carbon monoxide, sulfur oxides, nitrogen oxides, and lead. These pollutants can harm your health and the environment, and cause property damage. Of the six pollutants, particle pollution and ground-level ozone are the most widespread health threats ...

EPA calls these pollutants "criteria" air pollutants because it regulates them by developing human health-based and/or environmentally based criteria (science-based guidelines) for setting permissible levels. The set of limits based on human health is called *primary standards.* Another set of limits intended to prevent environmental and property damage is called *secondary standards.* A geographic area with air quality that is cleaner than the primary standard is called an *attainment area*; areas that do not meet the primary standard are called *nonattainment areas.*

Particle pollution, also known as *particulate matter* (PM), includes the very fine dust, soot, smoke, and droplets that are formed from chemical reactions, and are produced when fuels such as coal, wood, or oil are burned. For example, sulfur dioxide and nitrogen oxide gases from motor vehicles, electric power generation, and industrial facilities react with sunlight and water vapor to form particles. Particles may also come from fireplaces, wood stoves, unpaved roads, and crushing and grinding operations, and may be blown into the air by the wind.

Particles also cause haze reducing visibility in places like national parks and wilderness areas. In many parts of the United States, pollution has reduced the distance and clarity of what we see by 70%.

Fine particles can remain suspended in the air and travel long distances with the wind. Particles make buildings, statues, and other outdoor structures dirty. Trinity Church in downtown New York City was black until a few years ago, when cleaning off almost 200 years worth of soot brought the church's stone walls back to their original light pink color.

Before the 1990 Clean Air Act went into effect, EPA set limits on airborne particles smaller than 10 micrometers in diameter called PM10. In 1997, EPA published limits for fine particles, called PM2.5. To reduce particle levels, additional controls are being required on a variety of sources including power plants and diesel trucks.

3.5.1.2.2 Ground-Level Ozone

Ground-level ozone is a primary component of smog. Ground-level ozone can cause human health problems and damage forests and agricultural crops. Repeated exposure to ozone can make people more susceptible to respiratory infections and lung inflammation. It also can aggravate preexisting respiratory diseases, such as asthma. Active, healthy adults, such as construction workers, can experience a reduction in lung function and an increase in respiratory symptoms (chest pain and coughing) when exposed to low levels of ozone during periods of moderate exertion.

The main ingredients in forming ground-level ozone are called *volatile organic compounds* (VOCs) and nitrogen oxides (NOx). VOCs are released by cars burning gasoline and by petroleum refineries, chemical manufacturing plants, and other industrial facilities. The solvents used in paints and other consumer and business products contain VOCs. The 1990 Clean Air Act has resulted in changes in product formulas to reduce the VOC content of those products. Nitrogen oxides (NOx) are produced when cars and other sources like power plants and industrial boilers burn fuels such as gasoline, coal, or oil. The reddish-brown color you sometimes see when it is smoggy comes from the nitrogen oxides.

Weather and the lay of the land (for example, hills around a valley, high mountains between a big industrial city, and suburban or rural areas) help determine where ground-level ozone goes and how bad it gets. When temperature inversions occur (warm air stays trapped near the ground by a layer of cooler air) and winds are calm, high concentrations of ground-level ozone may persist for days at a time. As traffic and other sources add more ozone-forming pollutants to the air, the ground-level ozone gets worse.

3.5.1.2.3 Cars, Trucks, Buses, and Nonroad Equipment

Today, motor vehicles are responsible for nearly one-half of smog-forming VOCs, more than half of the NOx emissions, and about half of the toxic air pollutant emissions in the United States. Motor vehicles, including nonroad vehicles, now account for 75% of carbon monoxide emissions nationwide.

The total vehicle miles people travel in the United States increased 178% between 1970 and 2005 and continues to increase at a rate of two to three percent each year. In the United States, there are more than 210 million cars and light-duty trucks on the road. Beginning in the late 1980s, Americans began driving more vans, sport utility vehicles (SUVs), and pickup trucks. These bigger vehicles typically consume more gasoline per mile, and many of them pollute three to five times more than cars.

The Clean Air Act requires manufacturers to build cleaner engines, refiners to produce cleaner fuels, and certain areas with air pollution problems to adopt and run passenger vehicle inspection and maintenance programs. EPA has issued a series of regulations affecting passenger cars, diesel trucks, and buses, and so-called "nonroad" equipment (recreational vehicles, lawn and garden equipment, etc.) to reduce emissions as people buy new vehicles and equipment.

3.5.1.2.4 Cleaner Cars

A new car purchased today is well over 90% cleaner than a new vehicle purchased in 1970. As more of these cleaner vehicles enter the national fleet, harmful emissions should drop dramatically.

3.5.1.2.5 Lead and Other Toxic Pollutants

One of EPA's earliest accomplishments was the elimination of lead from gasoline. This effort was followed by even stronger restrictions on the use of lead in gasoline in the 1980s. In 1996, leaded gasoline was finally banned as a result of the Clean Air Act.

Under the Clean Air Act, EPA has also put into place standards to reduce toxic air emissions from mobile sources. These standards will cut toxic emissions from gasoline, vehicles, and even gas containers. Solutions include

Reformulated gasoline
Low-sulfur fuels
Alternative fuels

Alternative fuels are transportation fuels other than gasoline and diesel, including natural gas, propane, methanol, ethanol, electricity, and biodiesel. Renewable alternative fuels are made from biomass materials like wood, waste paper, grasses, vegetable oils, and corn. They are biodegradable and reduce carbon dioxide emissions. In addition, most alternative fuels are produced domestically, which is better for our economy and energy security, and helps offset the cost of imported oil.

The Clean Air Act also requires EPA to establish a national renewable fuel (RF) program. This program is designed to significantly increase the volume of renewable fuel that is blended into gasoline and diesel.

3.5.1.2.6 Cleaner Trucks, Buses, and "Nonroad" Equipment

- Heavy-duty trucks and buses account for about one-third of nitrogen oxides emissions and one-quarter of particle pollution emissions from transportation sources. In some large cities, the contribution is even greater. Similarly, nonroad diesel engines such as construction and agricultural equipment emit large quantities of harmful particle pollution and nitrogen oxides, which contribute to ground-level ozone and other pervasive air quality problems.
- EPA has issued rules to cut emissions from onroad and nonroad vehicles by more than 90%. Under the Clean Air Act, EPA is also addressing pollution from a range of nonroad sources, including locomotives and marine vessels, recreational vehicles, and lawn and garden equipment. Together these sources comprise a significant portion of emissions from the transportation sector.

3.5.1.3 Transportation Policies

- Transportation projects such as construction of highways and transit rail lines cannot be federally funded or approved unless they are consistent with state air quality goals. In addition, transportation projects must not cause or contribute to new violations of the air quality standards, worsen existing violations, or delay attainment of air quality standards.
- The conformity provisions require areas that have poor air quality now, or had it in the past, to examine the long-term air quality impacts of their transportation system and ensure that it is compatible with the area's clean air goals. In doing so, those areas must assess the impacts of growth on air pollution and decide how to manage growth. State and local agencies must work together to either change the transportation plan and/or the state air plan to achieve the necessary emission reductions.

3.5.1.3.1 Inspection and Maintenance Programs

- Proper maintenance of a car's engine and pollution control equipment is critical to reduce excessive air pollution. To help ensure that such maintenance occurs, the Clean Air Act requires certain areas with air pollution problems to run inspection and maintenance (I/M) programs. The 1990 Act also established the requirement that passenger vehicles be equipped with onboard diagnostics. The diagnostics system is designed to trigger a dashboard "check engine" light alerting the driver of a possible pollution control device malfunction. To help ensure that motorists respond to the "check engine" light in a timely manner, the act requires that I/M programs include an inspection of the onboard diagnostic system.

3.5.1.3.2 Interstate and International Air Pollution

- Air pollution does not recognize state or international boundaries. Pollutants can be carried long distances by the wind.
- States and tribes seeking to clean up air pollution are sometimes unable to meet EPA's national standards because of pollution blowing in from other areas. The Clean Air Act has a number of programs designed to reduce long-range transport of pollution from one area to another.
- The act gives any state or tribe the authority to ask EPA to set emission limits for specific sources of pollution in other (upwind) areas that significantly contribute to its air quality problems. States and tribes can petition EPA to require the upwind areas to reduce air pollution.
- The act provides for interstate commissions to develop regional strategies for cleaning up air pollution. For instance, state and tribal governments from Maine to Virginia, the government of the District of Columbia, and EPA are working together through the Ozone Transport Commission (OTC) to reduce ground-level ozone along the east coast.

3.5.1.3.3 Reducing Acid Rain

Acid rain and other forms of acid precipitation such as acid snow, acid fog or mist, or dry forms of acidic pollution such as acid gas and acid dust can be formed in the atmosphere and fall to Earth causing human health problems, hazy skies, environmental problems, and property damage. Acid precipitation is produced when certain types of air pollutants mix with the moisture in the air to form an acid. These acids then fall to Earth as rain, snow, or fog. Even when the weather is dry, acid pollutants may fall to Earth in gases or particles.

Sulfur dioxide (SO_2) and nitrogen oxides (NOx) are the principal pollutants that cause acid precipitation. SO_2 and NOx emissions released to the air react with water vapor and other chemicals to form acids that fall back to Earth. Power plants burning coal and heavy oil produce over two-thirds of the annual SO_2 emissions in the United States. The majority of NOx (about 50%) comes from cars, buses, trucks, and other forms of transportation. About 40% of NOx emissions are from power plants. The rest is emitted from various sources like industrial and commercial boilers.

High levels of SO_2 in the air aggravate various lung problems in people with asthma and can cause breathing difficulties in children and the elderly. In some instances, breathing high levels of SO_2 can even damage lung tissue and cause premature death.

The 1990 changes to the Clean Air Act use a market-based cap and trade approach, the program sets a permanent cap on the total amount of SO_2 that may be emitted by electric power plants nationwide.

Each allowance is worth one ton of SO_2 emissions released from the plant's smokestack. Plants may only release the amount of SO_2 equal to the allowances they have been issued. If a plant expects to release more SO_2 than it has allowances, it has to purchase more allowances or use technology and other methods to control emissions. A plant can buy allowances from another power plant that has more allowances than it needs to cover its emissions.

There is an allowances market that operates like the stock market, in which brokers or anyone who wants to take part in buying or selling allowances can participate. Allowances are traded and sold nationwide.

All power plants covered by the Acid Rain Program have to install continuous emission monitoring systems, and instruments that keep track of how much SO_2 and NOx the plant's individual units are releasing. Power plant operators keep track of this information hourly and report it electronically to EPA four times each year.

3.5.1.4 Protecting the Stratospheric Ozone Layer

- Ozone can be good or bad depending on where it is located. Close to the Earth's surface, ground-level ozone is a harmful air pollutant. Ozone in the stratosphere, high above the Earth, protects human health and the environment from the sun's harmful ultraviolet radiation. This natural shield has been gradually depleted by man-made chemicals. So in 1990, Congress added provisions to the Clean Air Act for protecting the stratospheric ozone layer.
- Ozone in the stratosphere, a layer of the atmosphere located 10 to 30 miles above the Earth, serves as a shield, protecting people and the environment from the sun's harmful ultraviolet radiation. The stratospheric ozone layer filters out harmful sun rays, including a type of sunlight called ultraviolet B. Exposure to ultraviolet B (UVB) has been linked to cataracts (eye damage) and skin cancer. Scientists have also linked increased UVB exposures to crop injury and damage to ocean plant life.
- In the mid-1970s, scientists became concerned that chlorofluorocarbons (CFCs) could destroy stratospheric ozone. At that time, CFCs were widely used as aerosol propellants in consumer products such as hairsprays and deodorants, and as coolants in refrigerators and air conditioners. In 1978, the U.S. government banned CFCs as propellants in most aerosol uses.
- Scientists have been monitoring the stratospheric ozone layer since the 1970s. In the 1980s, scientists began accumulating evidence that the ozone layer was being depleted. The ozone

hole in the region of the South Pole, which has appeared each year during the Antarctic winter (our summer), often is bigger than the continental United States. Between 1978 and 1997, scientists measured a 5% loss of stratospheric ozone—a significant amount.

- Over 190 countries, including the major industrialized nations such as the United States, have signed the 1987 Montreal Protocol, which calls for elimination of chemicals that destroy stratospheric ozone. Countries that signed the protocol are committed to limiting the production and use of those chemicals.
- The 1990 Clean Air Act required EPA to set up a program for phasing out production and use of ozone-destroying chemicals. In 1996, U.S. production ended for many of the chemicals capable of doing the most serious harm such as CFCs, halons, and methyl chloroform.
- Unfortunately, it will be about 60 years before the stratospheric ozone layer heals. Because of the ozone-destroying chemicals already in the stratosphere and those that will arrive within the next few years, stratospheric ozone destruction will likely continue throughout the decade. September 24, 2006, tied for the largest ozone hole on record at 29 million square kilometers (11.4 million square miles). The year 2006 also saw the second largest sustained ozone hole.
- The Clean Air Act includes other steps to protect the ozone layer. The act encourages the development of "ozone-friendly" substitutes for ozone-destroying chemicals. Many products and processes have been reformulated to be more "ozone-friendly." For instance, refrigerators no longer use CFCs.
- Sometimes it isn't easy to phase out an ozone-destroying chemical. For instance, substitutes have not been found for CFCs used in certain medical applications. The limit on the production of methyl bromide, a pesticide, was extended because farmers did not yet have an effective alternative. Despite the inevitable delays because of technical and economic concerns, ozone-destroying chemicals are being phased out, and, with continued work, over time the protective ozone layer will be repaired.

3.5.1.5 Permits and Enforcement

3.5.1.5.1 Permits

- The Clean Air Act in 1990 includes an operating permit program for larger industrial and commercial sources that release pollutants into the air. Operating permits include information on which pollutants are being released, how much may be released, and what kinds of steps the source's owner or operator is required to take to reduce the pollution.
- Operating permits are especially useful for businesses covered by more than one part of the Clean Air Act and additional state or local requirements, since information about all of a source's air pollution is in one place. The permit program simplifies and clarifies businesses' obligations for cleaning up air pollution and can reduce paperwork.
- Businesses seeking permits have to pay permit fees, much like car owners paying for car registrations. These fees pay for the air pollution control activities related to operating permits.

3.5.1.5.2 Enforcement

- The Clean Air Act gives EPA important enforcement powers. The 1990 amendments strengthened EPA's power to enforce the act, increasing the range of civil and criminal sanctions available. In general, when EPA finds that a violation has occurred, the agency can issue an order requiring the violator to comply, issue an administrative penalty order (use EPA administrative authority to force payment of a penalty), or bring a civil judicial action (sue the violator in court).

3.5.1.6 Public Participation

Public participation is a very important part of the 1990 Clean Air Act. Throughout the act, different provisions give the public opportunities to take part in determining how the law is carried out. When EPA is working on a major rule, the agency will hold hearings in various cities across the country, at which the public can comment. You can also submit written comments directly to EPA for inclusion in the public record associated with that rule. Or, for instance, you can participate in development of a state or tribal implementation plan.

The 1990 Clean Air Act gives you opportunities to take direct action to get pollution cleaned up in your community. You can get involved in reviewing air pollution permits for industrial sources in your area. You also can ask EPA, your state or tribe to take action against a polluter, and, in some cases, you may be able to take legal action against a source's owner or operator.

3.5.2 Edited Material From the EPA About the Clean Water Act

The 1972 amendments to the Federal Water Pollution Control Act (known as the Clean Water Act or CWA) provide the statutory basis for the NPDES permit program and the basic structure for regulating the discharge of pollutants from point sources to waters of the United States. Section 402 of the CWA specifically required EPA to develop and implement the NPDES Permit Program.

The CWA gives EPA the authority to set effluent limits on an industry-wide (technology-based) basis and on a water-quality basis that ensure protection of the receiving water. The CWA requires anyone who wants to discharge pollutants to first obtain an NPDES permit, or else that discharge will be considered illegal.

The CWA allowed EPA to authorize the NPDES Permit Program to state governments, enabling states to perform many of the permitting, administrative, and enforcement aspects of the NPDES Permit Program. In states that have been authorized to implement CWA programs, EPA still retains oversight responsibilities.

The key sections of the CWA that directly relate to the NPDES Permit Program include:

Title I—Research and Related Programs
  Section 101—Declaration of Goals and Policy
Title II—Grants for the Construction of Treatment Works
Title III—Standards and Enforcement
  Section 301—Effluent Standards
  Section 302—Water Quality-Related Effluent Limitations
  Section 303—Water Quality Standards and Implementation Plans
  Section 304—Information and Guidelines [Effluent]
  Section 305—Water Quality Inventory
  Section 307—Toxic and Pretreatment Effluent Standards
Title IV—Permits and Licenses
  Section 402—National Pollutant Discharge Elimination System
  Section 405—Disposal of Sewage Sludge

Title V—General Provisions
Section 510—State Authority
Section 518—Indian Tribes

3.5.3 Comprehensive Environmental Response, Compensation, and Liability Act (CERCLA) Hazardous Substances

CERCLA hazardous substances are substances that are considered severely harmful to human health and the environment. Many are commonly used substances which are harmless in their normal uses, but are quite dangerous when released. They are defined in terms of those substances either specifically designated as hazardous under the Comprehensive Environmental Response, Compensation, and Liability Act (CERCLA), commonly known as the Superfund law, or those substances identified under other laws. In all, the Superfund law designates more than 800 substances as hazardous and identifies many more as potentially hazardous due to their characteristics and the circumstances of their release.

Superfund's definition of a hazardous substance includes the following:

- Any element, compound, mixture, solution, or substance designated as hazardous under Section 102 of CERCLA.
- Any hazardous substance designated under Section 311(b)(2)(a) of the Clean Water Act (CWA), or any toxic pollutant listed under Section 307(a) of the CWA. There are over 400 substances designated as either hazardous or toxic under the CWA.
- Any hazardous waste having the characteristics identified or listed under Section 3001 of the Resource Conservation and Recovery Act.
- Any hazardous air pollutant listed under Section 112 of the Clean Air Act (CAA), as amended. There are over 200 substances listed as hazardous air pollutants under the Clean Air Act.
- Any imminently hazardous chemical substance or mixture which the EPA Administrator has "taken action under" Section 7 of the Toxic Substances Control Act.

Hazardous waste is defined under the Resource Conservation and Recovery Act (RCRA) as a solid waste (or combination of solid wastes) which, because of its quantity, concentration, or physical, chemical, or infectious characteristics, may (1) cause or contribute to an increase in mortality or an increase in serious irreversible, or incapacitating illness or (2) pose a substantial present or potential hazard to human health or the environment when improperly treated, stored, transported, disposed of, or otherwise managed. In addition, under RCRA, EPA establishes four characteristics that will determine whether a substance is considered hazardous, including ignitability, corrosiveness, reactivity, and toxicity. Any solid waste that exhibits one or more of these characteristics is classified as a hazardous waste under RCRA and, in turn, as a hazardous substance under Superfund.

The terms *hazardous substance* and *pollutant* or *contaminant* do not include petroleum or natural gas. EPA conducts emergency responses to incidents involving petroleum and nonpetroleum oils separately from its responses to hazardous substance incidents. Throughout the Emergency Response Program, the term *hazardous substance* includes pollutants and contaminants.

In addition to the hazardous substances identified under the Superfund law, the Title III amendments to Superfund, also known as the Emergency Planning and Community Right-to-Know Act (EPCRA), identify several hundred hazardous substances for their extremely toxic properties. EPA designated them as "extremely hazardous substances" to help focus initial chemical emergency response planning efforts.

Hazardous waste is waste that is dangerous or potentially harmful to our health or the environment. Hazardous wastes can be liquids, solids, gases, or sludges. They can be discarded commercial products, like cleaning fluids or pesticides, or the by-products of manufacturing processes.

More about hazardous waste and the regulations that govern it:

- **Definition of Solid Waste (DSW)**: Before a material can be classified as a hazardous waste, it must first be a solid waste as defined under RCRA. Resources, including an interactive tool, are available to help.
- **Hazardous wastes** are divided into listed wastes, characteristic wastes, universal wastes, and mixed wastes. Specific procedures determine how waste is identified, classified, listed, and delisted.
- **Generators**: Hazardous waste generators are divided into categories based on the amount of waste they produce each month. Different regulations apply to each generator category.
- **Transporters**: Hazardous waste transporters move waste from one site to another by highway, rail, water, or air. Federal and, in some cases, state regulations govern hazardous waste transportation, including the Manifest System.
- **Treatment, Storage, and Disposal (TSD)**: Requirements for TSD facilities govern the treatment, storage, and disposal of hazardous waste, including land disposal, the permitting process and requirements for TSD facilities.
- **Waste Minimization**: EPA, states, and industries are working to reduce the amount, toxicity, and persistence of wastes that are generated.
- **Hazardous Waste Recycling**: EPA is addressing safe and protective reuse and reclamation of hazardous materials.
- **Corrective Action**: RCRA compels those responsible for releasing hazardous pollutants into the soil, water, or air to clean up those releases.
- **Test Methods**: EPA has a variety of analytical chemistry and characteristic testing methodologies, environmental sampling and monitoring, and quality assurance in place to support RCRA.
- **International Waste**: EPA provides information and guidance on regulations, agreements, initiatives, and other developments in waste policy and law, both in the United States and abroad.

3.5.4 Emergency Planning and Community Right-to-Know Act (EPCRA) Requirements

The Emergency Planning and Community Right-to-Know Act (EPCRA) of 1986 was created to help communities plan for emergencies involving hazardous substances. The act establishes requirements for federal, state, and local governments, Indian tribes, and industry regarding emergency planning and "community right-to-know" reporting on hazardous and toxic chemicals. The community right-to-know provisions help increase the public's knowledge and access to information on chemicals at individual facilities, their uses, and releases into the environment. States and communities, working with facilities, can use the information to improve chemical safety and protect public health and the environment.

There are four major provisions of EPCRA:

- Emergency Planning (Sections 301–303)

- Emergency Release Notification (Section 304)
- Hazardous Chemical Storage Reporting (Sections 311–312)
- Toxic Chemical Release Inventory (Section 313)

3.5.4.1 EPCRA Requirements

3.5.4.1.1 Local Emergency Planning Requirements

EPCRA local emergency planning requirements (Sections 301 to 303) stipulate that every community in the United States must be part of a comprehensive emergency response plan. Facilities are required to participate in the planning process.

- State Emergency Response Commissions (SERCs) oversee the implementation of EPCRA requirements in each state.
- Local Emergency Planning Committees (LEPCs) work to understand chemical hazards in the community, develop emergency plans in case of an accidental release, and look for ways to prevent chemical accidents. LEPCs are made up of emergency management agencies, responders, industry, and the public.

3.5.4.1.2 Chemical Reporting Requirements

According to the EPCRA chemical reporting requirements, facilities must report the storage, use, and release of certain hazardous chemicals.

3.5.4.1.2.1 Proposed Revisions to the Emergency and Hazardous Chemical Inventory Forms (Tier I and Tier II)

On August 8, 2011, EPA proposed revisions to the Emergency and Hazardous Chemical Inventory Forms under Section 312 of the Emergency Planning and Community Right-to-Know Act (EPCRA) to add new data elements and revise some existing data elements. The proposed changes are intended to meet the purpose of EPCRA, which is "… to encourage and support state and local planning for emergencies caused by the release of hazardous chemicals and to provide citizens and governments with information concerning potential chemical hazards present in their communities."

The proposed revisions:

- Respond to stakeholder requests. EPA is proposing to add new data elements to the Tier I and Tier II forms in an effort to make the forms more useful for state, local, and tribal agencies.
- Make reporting easier for facilities.
- Are intended to provide clarity in reporting while maintaining protection of human health and the environment.
- May impose minimal reporting burden on facilities since the data elements proposed are readily available to the facility. Revising the existing data elements will make the forms more user-friendly and ease reporting requirements for facilities.

Organizations and facilities subject to Section 312 of EPCRA and its implementing regulations found in 40 CFR 370 may be affected by this rule.

3.5.4.1.2.2 Amendments to Rule for Extremely Hazardous Substances (EHS) and Threshold Planning Quantities (TPQs) for Solids in Solutions

On April 15, 2011, the U.S. Environmental Protection Agency (EPA) proposed amendments to revise the way the regulated community applies the threshold planning quantities (TPQs) for Extremely Hazardous Substances (EHS). This applies to EHS that are nonreactive solid chemicals in solution form. With these amendments, EPA proposes modifying the assumptions used to develop the TPQs for solid EHS chemicals in solution. EPA is proposing these amendments because available data shows less potential for the solid chemical in solution to remain airborne in the event of an accidental release.

There are 157 EHS chemicals which could potentially be affected by this change. The affected chemicals are identified in Appendix C of the Technical Support Document for Revised TPQ Method for EHS Solids in Solution, which is the docket to this rule. These 157 chemicals also appear with two TPQs (the higher TPQ is 10,000 pounds) in Appendices A and B of 40 CFR part 355.

Organizations and facilities subject to Section 302 of the Emergency Planning and Community Right-to-Know Act (EPCRA) and its implementing regulations found in 40 CFR 355 subpart B may be affected by this rule.

3.5.4.1.2.3 Reporting Options for Sections 311 and 312 and Interpretations

The U.S. Environmental Protection Agency (EPA) provided draft guidance in the preamble to the June 8, 1998, proposed rule (63 FR 31268) to streamline the reporting requirements for facilities under Sections 311 and 312 of the Emergency Planning and Community Right-to-Know Act of 1986 (EPCRA). The agency did not propose any regulatory changes, but sought comments on the reporting options. EPA is now providing guidance on the reporting options. The objective for this guidance is also to provide state and local agencies with flexibility in implementing Sections 311 and 312 of EPCRA.

3.5.4.1.2.4 Administrative Reporting Exemption for Air Releases of Hazardous Substances From Animal Waste at Farms

The U.S. Environmental Protection Agency (EPA) is announcing a final rule to provide an administrative reporting exemption for releases to the air from animal waste at farms of any hazardous substance at or above the reportable quantity for those hazardous substances. EPA is saying that these reports are unnecessary because there is no reasonable expectation that a federal response would be made as a result of such reports. The final rule reduces the burden of complying with Comprehensive Environmental Response, Compensation, and Liability Act (CERCLA) and to a limited extent, the Emergency Planning and Community Right-to-Know Act (EPCRA) reporting requirements on the regulated community. This rule is effective January 20, 2009.

3.5.4.1.2.5 Final Amendments to EPCRA Regulations

EPA finalized several changes to the Emergency Planning (Section 302), Emergency Release Notification (Section 304), and Hazardous Chemical Reporting (Sections 311 and 312) regulations that were proposed on June 8, 1998 (63 FR 31268). These changes include clarification on how to report hazardous chemicals in mixtures, and changes to Tier I and Tier II forms. Additionally, the rules now use a question and answer format. Facilities subject to EPCRA reporting, state emergency response commissions (SERCs), local emergency planning committees (LEPCs), and fire departments should become familiar with the new regulation. The final rule was effective on December 3, 2008.

3.5.5 Summary of the Resource Conservation and Recovery Act

This can be best summarized at the appropriate level here by looking at the original EPA press release from 1976, titled: "New Law to Control Hazardous Wastes, End Open Dumping, Promote Conservation of Resources":

> Under a new law enacted October 21, 1976, the handling and disposal of hazardous wastes, which are generated mainly by industry, will come under federal/state regulation. The law also requires that open dumping of all solid wastes be brought to an end throughout the country by 1983.
>
> The Resource Conservation and Recovery Act (P.L. 94-580) also calls for research, demonstrations, studies, training, information dissemination, and public participation activities to enlarge the base of knowledge and public involvement necessary for developing strong state and local programs.
>
> Partly as a result of pollution controls that keep wastes out of the air and water, growing amounts of solid wastes are being generated and deposited on the land. Disposal on land has gone largely uncontrolled, resulting in numerous instances of serious effects on human health and environmental quality. The contamination of groundwaters by substances leaching from disposal sites is a primary concern. The most urgent objective of the new law is to prevent this and other environmental effects of improper disposal.
>
> In signing the law, President Ford cited the special threat in hazardous waste disposal, calling it "one of the highest priority environmental problems confronting the Nation." Under the law, EPA is required to identify and publish a list of hazardous wastes within 18 months and to set standards for the handling, transportation, and ultimate disposal of these wastes. Under guidelines to be developed by EPA, States are to establish regulatory programs; if States fail to do so, EPA regulations will apply.
>
> Civil and criminal penalties are established for violation—up to $25,000 per day of noncompliance, a year in prison, or both.
>
> To implement the open dumping prohibition, EPA is directed to establish criteria for identifying open dumps and for identifying sanitary landfills no later than October 1977, and the agency will conduct a national inventory of all open dumps within the 12 months that follow. Special grant assistance to help meet new requirements for land disposal facilities will be available for rural communities.

Other major provisions of the law include:

- A requirement that all federal procurement agencies procure items composed of the maximum allowable percentage of recycled materials.
- A requirement that public participation must be promoted in the development of all federal and state regulations, guidelines, information, and programs under the act.
- Permission for citizens to bring suits to obtain compliance with the law.
- Requirement of a number of special studies, in areas such as sewage sludge management, low-technology means of resource recovery, measures to reduce the generation of waste, waste collection practices, management of mining and agricultural wastes, and economic incentives to promote recycling and waste reduction.

3.6 29 CFR

We are only interested in a very small part of the 29 CFR. The entire title is currently divided into 40 chapters, only 14 of which have been written and codified. There are a few specific areas that are critical to appropriate corporate or strategic management for life-cycle management of hazardous materials.

The first is § 1910.120 HAZARDOUS WASTE OPERATIONS AND EMERGENCY RESPONSE. Some key concepts here starting with a 29 CFR definition: "*Hazardous waste site* or *site* means any facility or location within the scope of this standard at which hazardous waste operations take place" (29 CFR 2011). That means any commercial activity where anything labeled as a hazardous waste, understand that includes fluorescent lights batteries and waste lubrication oil, is covered by the available regulation. This section specifically creates a statutory requirement for a formal training program; there are other elements as well. Many have heard the term "HAZWOPER" training; this is where the term and the requirement are created and defined. Courses are federally mandated, and they require specific training; this has become a very lucrative business for a number of providers. Unfortunately, it has also caused some shifting among most of the corporate- or business-focused hazardous wastes operations as distinct from hazardous material management. A cost requirement has arisen out of OSHA. There also are far too many trainers providing fully compliant training that fails to recognize the underlying reason we have regulations and the larger issues of life cycle management. We have no argument about the legitimacy or the need for this. Today, however, the problem is that far too many people see this as the entire focus of hazardous materials management rather than recognizing that this regulation is addressing a reactive requirement rather than a proactive elimination mitigation strategy.

The second is § 1910.1200 HAZARD COMMUNICATION. Probably the best regulatory, authority, definition, requirement, and program in the entire 29 CFR. At this point there are a handful of core concepts one should understand. The first part of this very long and complex section simply says one must have a hazard communication program, and that every employee, including remote site workers and field employees, who might come in contact with any such items must have a certain amount of training. They must be able to access the MSDS at their location in a form/format that is easily accessible and usable; downstream users/customers must be provided with copies of the MSDS. There are, in part two or three, other specific requirements worth noting in terms of hazard communication. Foreign producers are not required to create an MSDS sheet, however, the importing organization, including governments and businesses are required to generate an MSDS for such imported material and assure that the MSDS is distributed with the material throughout the rest of the distribution system from the point of import onward. Any material moved or used within the United States that represents any sort of environmental health and safety hazard requires an MSDS and that MSDS must be available at multiple points in the transportation distribution and warehouse system all the way to the retail user or for businesses in the appropriate workspaces. There continue to be conflicting interpretations as to whether availability in a workspace includes electronic availability rather than hard-copy access. In many situations where there is an emergent need, there will also be a loss of power communication, so while one may make a reasonable argument for allowing the MSDS to be available electronically there are a large number of factors that mitigate against such a decision. Electronic backup at the site is certainly advisable at the right place. The primary approach should be its organization in a way that the local workers can access the materials. For some organizations, that may mean multiple copies, although possibly organized differently, in different workspaces within the same physical facility. As an example, there might be a need in the shipping receiving area, as well as a need for "shop stores" on the production floor where the hazardous material is actually used/applied. What is overlooked or misunderstood is that putting together one large "book" with the full MSDS and then placing copies at different locations within this facility does not meet the very specific language and intent of this section of the regulation.

The next is § 1910.1201 RETENTION OF DOT MARKINGS, PLACARDS AND LABELS. The title here is self-explanatory. The point that needs to be reinforced is that transportation, for multiple reasons, is the "lowest common denominator," and, the universally/globally accepted transportation classification system is the most widely used, understood, and recognized system for identifying dangerous goods/hazardous materials. A thorough understanding of the classification system allows one to see that symbols or pictograms are all that is needed to broadly identify all materials, therefore understanding of any particular language is not critical, although being able to recognize Arabic numbers is important.

The third is § 1960 BASIC PROGRAM ELEMENTS FOR FEDERAL EMPLOYEE OCCUPATIONAL SAFETY AND HEALTH PROGRAMS AND RELATED MATTERS. There are two reasons this section is included: the first is to demonstrate that these regulations apply to government. While it is not clearly stated, that means at all levels from the national level down to the lowest jurisdiction and of course including tribal governments at all levels, and to give you some examples of how to construct a program in the broadest terms and who it must include. To the author, the "who" remains one of the largest problems, as far too many regulations focus on frontline users and perhaps first-line supervisors, not the entire management and supervisory chain. In any organization that sees hazardous materials pass into and out of its workspaces, it is safe to assume that greater than 90% of the employees do, in fact, fall under the definition of HAZMAT employee as defined in 49 CFR.

3.7 10 CFR, 13 CFR, 23 CFR, 46 CFR, and Other Applicable U.S. Laws/Regulations

The material to follow is initially arranged in ascending order of the applicable CFR number as an easy and hopefully unbiased approach to a very rich but unfortunately ambiguous area.

3.7.1 10 CFR Energy

There are two areas within the 10 CFR that directly impact the life cycle management of all hazardous materials if one is accepting, not necessarily agreeing with, the concept of airborne emissions, as well as the more traditionally recognized liquid and solid streams of materials. The Department of Energy has overall responsibility for the writing and promulgation of the regulation and shares enforcement responsibilities with many other agencies including multiple modal administrations within the DOT. Part 171 deals with the transportation of radioactive material. When you realize that such materials occur in commercial laboratory test instrumentation—as sources for atomic emission and absorption testing involving metal and other building materials as an example—and in media used for all sorts of medical tests and doctors' offices as well as hospitals, you begin to recognize that a much larger segment of the population does have to be aware of the regulations than one might originally think. Technically, many smoke detectors, as well as sights on commercial and military weapons, also contain radioactive material but at the current time, the governments of the world have chosen not to regulate this at the retail level.

In 2007 the Congress of the United States passed the Energy Independence and Security Act. As with most major legislation, the impacts are spread out over the entire spectrum of society and over extended periods of time. There are opportunities for business, as well as new requirements. As a tertiary example, this is where the wholesale replacement of incandescent light bulbs with CFLs comes from. Requirements for both commercial users and builders of consumer electronics and appliances are also driven by the EISA; you will find most of that addressed in 10 CFR within the broad area of Chapter II subchapter D. The layout of the key related portions of the 10 CFR follow below:

CHAPTER I—NUCLEAR REGULATORY COMMISSION
71.0 to 71.137 PACKAGING AND TRANSPORTATION OF RADIOACTIVE MATERIAL

CHAPTER II—DEPARTMENT OF ENERGY
SUBCHAPTER B—CLIMATE CHANGE
SUBCHAPTER C—[RESERVED]
SUBCHAPTER D—ENERGY CONSERVATION

3.7.2 13 CFR Commerce

The 13 CFR deals with economic development from a hazardous material life cycle management standpoint. Most of the positives have to do with special programs, special incentives, and unique opportunities; the relevant information falls within Chapter III. Realistically, 13 CFR is something one would look at as part of an expansion or new facility exploration, as distinct from looking for opportunities in existing facilities. We will not spend any more time on this particular CFR, but when used in conjunction with 23 CFR and the EISA, portions of 10 CFR companies can find ways to create leverage advantage while gaining global recognition for environmental stewardship. The layout of the key-related portions of the 13 CFR follow below:

TITLE 13—BUSINESS CREDIT AND ASSISTANCE
CHAPTER I—SMALL BUSINESS ADMINISTRATION
CHAPTER III—ECONOMIC DEVELOPMENT ADMINISTRATION, DEPARTMENT OF COMMERCE

3.7.3 23 CFR Highways

This is a big one; it represents a very complex set of relationships and equations. The name for this title is a little bit misleading; the fact is that it covers an area that is extremely important to communities and individual business entities, but it is really driven by the EPA, which adds to the confusion and complexity. The area we are interested in here is a direct result of the Clean Air Act. In an attempt to keep this fairly simple, one needs to understand that the Clean Air Act focuses very heavily on greenhouse gases, nonpoint source pollution, transportation versus population density, urban/regional planning, transit, environmental justice, and a number of other related areas. In the end, the Clean Air Act, through 23 CFR, drives very large amounts of all of our community planning and zoning. The relationship between air quality and the primary sources of air pollution is why this is such a critical regulation for hazardous material life cycle management. The statutory basis that all need to be aware of initially is Congestion Mitigation and Air Quality Improvement or CMAQ. There are other areas in other regulations that help tie this in, but you will find that 23 CFR chapter 1 subchapter E part 450 contains a lot of the relevant information. Understand that this directly impacts on every workplace without regard to whether it is a governmental workplace, a business workplace, whether it is charitable or for profit, or qualifies as an NGO. The trade-off between mass transit and privately owned vehicles is one part of it; the trade-off of effectively using marine ports and rails is another part of it. Any and all discharges from individual business entities, as well as their impact on municipal water treatment and commercial power generation, all tie together. The layout of the key related portions of the 23 CFR follow:

TITLE 23—HIGHWAYS
CHAPTER I—FEDERAL HIGHWAY ADMINISTRATION, DEPARTMENT OF TRANSPORTATION
SUBCHAPTER E—PLANNING AND RESEARCH

420 **420.101 to 420.209** PLANNING & RESEARCH PROGRAM ADMINISTRATION
450 **450.100 to 450.338** PLANNING ASSISTANCE AND STANDARDS
460 **460.1 to 460.3** PUBLIC ROAD MILEAGE FOR APPORTIONMENT OF HIGHWAY SAFETY FUNDS
470 **470.101 to 470.115** HIGHWAY SYSTEMS

3.7.4 33 CFR Navigation and Navigable Waters

The title here is, again, misleading. Much in this CFR is driven by the Clean Water Act and has to do with oil pollution. Conceptually, one must understand that a basic assumption here is that any form of liquid pollution that is discharged and permeates into the soil will eventually reach both drinking water and navigable water, therefore this regulation is yet another that plays a crucial role in the overall life cycle management of hazardous materials. As a sidelight to trying to reinforce why this is so important, it is worth repeating that very many of the United States' largest 400 metropolitan areas that are located along the coasts had for decades barged their solid waste out into the ocean and dumped it. We are not going to debate whether that was right or wrong but today science proves that this was not a good idea. Those previous actions have allowed the federal government to get involved in the hazardous material life cycle management through the 33 CFR.

There are abbreviated tables of contents of the applicable parts of the previously noted regulations provided in the appendices. All of these have less direct impact than the three primary regulations, and it would have become extremely cumbersome to try to place too much material from any of them into this text as their application is driven so heavily by what others in the immediate vicinity are doing or have done, and what population densities are. In the simplest terms, for CMAQ in nonattainment areas, anything and everything—which includes adding more jobs, which means more commuters—has direct impact on both the amount of federal funding available and the qualifying standards for states and local jurisdictions to be able to obtain that funding. That is why every aspect of life cycle management of hazardous material becomes an important consideration for a company that looks toward growing or moving into a new facility or adding additional facilities in other locations. The layout of the key related portions of the 33 CFR follow:

TITLE 33—NAVIGATION AND NAVIGABLE WATERS
CHAPTER I—COAST GUARD, DEPARTMENT OF HOMELAND SECURITY
SUBCHAPTER L—WATERFRONT FACILITIES
SUBCHAPTER M—MARINE POLLUTION FINANCIAL RESPONSIBILITY AND COMPENSATION
SUBCHAPTER N—OUTER CONTINENTAL SHELF ACTIVITIES
SUBCHAPTER O—POLLUTION

3.8 NIOSH, OSHA, EPA, PHMSA, MSA, and Other Specific Federal Organizations Involved

In addition to the three primary organizations that are directly involved in life cycle management of hazardous material (DOT, EPA OSHA), there are an extremely large number of additional specific entities involved, spread out among all the cabinet level functions of the federal government. Many are mirrored at the state, regional, or local level, as well. This section merely highlights a number of more important organizations. Depending on where you are, what the primary pollution-related areas are in your immediate region, and what population densities are, any or all of the following organizations might be an important stakeholder for your operation. Let me repeat yet again that all executive branches of government are subject to regulation. Based on the language in the U.S. Code, there is a gray area that may be interpreted in a number of different ways. However, in the hazardous materials transportation area—and in the environmental and occupational safety and health areas—precedents have already been set and case law already exists. One may argue that in terms of civil and criminal penalties, government agencies may not always be able to hold government agencies responsible. What is clear and what has been demonstrated time and time again is that in those cases where it was judged inappropriate for one agency to hold another agency financially responsible, it is *not* unusual or improper for the prosecuting agency to hold the "employees" individually responsible.

The following is an abbreviated list of some of the more prominent federal organizations that could very well have direct impacts on a business entity that purchases materials, produces a product, and sells a product. These would apply, to a much lesser extent, with service-providing organizations:

AGRICULTURE: APHIS, FDA
DEFENSE: COE, NGB, CSTs
HEALTH AND HUMAN SERVICES: CDC, NIOSH
HOMELAND SECURITY: FEMA, USCG
INTERIOR: BLM, MSA
LABOR: OSHA
TRANSPORTATION: FAA, FRA, FHWA, MARAD, FMCSA, PHMSA

3.9 ANSI, ISO, NFPA, ACGIH, NIST, and Other Standard-Setting Organizations

There will always be a need for consensus standards and statutory requirements. Some segments of society or business self-regulate because it is just too inherently, intrinsically dangerous not to self-regulate. In general, the airline industry falls into that category and, in fact, the "working" document for shipping dangerous goods worldwide is the industry consensus standard. Because of differences in actual publication date, there are times when this publication will be more stringent than the statutory requirement, although, typically, the two match up pretty well. It is not unusual for a U.S. consensus standard, such as the IATA dangerous goods regulations, to be more restrictive than a statutory requirement. Another excellent example is the ACGIH and its TLVs and PEIs. The U.S. government requires this data to be included on MSDS sheets, but the government does not develop the data or incorporate a manager to make it statutory. There is one federal organization deeply involved in the area of standards and a number of professional or trade organizations deeply involved in standards setting. Much of what is accepted by government comes from external sources. Today, there are a number of organizations whose standards are critical to life cycle management of hazardous materials. All of those organizations are listed below simply because there are too many standards, and so many are only applicable to certain segments or types of businesses that would be impossible to either provide information from the standards or even list all the organizations and their standards. The following list shows government organization at the top and then lists the other key organizations alphabetically:

National Institute of Standards and Technology
American Conference of Governmental Industrial Hygienists (ACGIH)
American National Standards Institute (ANSI)
American Petroleum Institute
ASTM—International Standards Worldwide
Compressed Gas Association
ISO—International Organization for Standardization
National Fire Protection Association
The Chlorine Institute

A quick look at key data requirements for the 16 parts of the ANSI Z 400 MSDS consensus standard:

SECTION 1: PRODUCT AND COMPANY IDENTIFICATION

PRODUCT NAME:
SYNONYMS:
PRODUCT CODES:

EMERGENCY PHONE:
CHEMTREC PHONE:
OTHER CALLS:
FAX PHONE:

CHEMICAL NAME:
CHEMICAL FAMILY:
CHEMICAL FORMULA:

PRODUCT USE:
PREPARED BY:

SECTION 2: COMPOSITION/INFORMATION ON INGREDIENTS

INGREDIENT:

CAS NO.	% WT	% VOL
	ppm	mg/m3
OSHA PEL-TWA:		
OSHA PEL STEL:		
OSHA PEL CEILING:		
ACGIH TLV-TWA:		
ACGIH TLV STEL:		
ACGIH TLV CEILING:		

SECTION 3: HAZARDS IDENTIFICATION EMERGENCY OVERVIEW:

ROUTES OF ENTRY:
POTENTIAL HEALTH EFFECTS
EYES:
SKIN:
INGESTION:
INHALATION:
ACUTE HEALTH HAZARDS:
CHRONIC HEALTH HAZARDS:
MEDICAL CONDITIONS GENERALLY AGGRAVATED BY EXPOSURE:
CARCINOGENICITY
OSHA: ACGIH: NTP: IARC:
OTHER:

SECTION 4: FIRST AID MEASURES

EYES:
SKIN:
INGESTION:
INHALATION:
NOTES TO PHYSICIANS OR FIRST AID PROVIDER:
SECTION 4 NOTES:

SECTION 5: FIRE-FIGHTING MEASURES

FLAMMABLE LIMITS IN AIR, UPPER:
(% BY VOLUME) LOWER:
FLASH POINT:
F:
C:
METHOD USED:
AUTOIGNITION TEMPERATURE:
F:
C:

NFPA HAZARD CLASSIFICATION
HEALTH: FLAMMABILITY: REACTIVITY:
OTHER:
HMIS HAZARD CLASSIFICATION
HEALTH: FLAMMABILITY: REACTIVITY:
PROTECTION:
EXTINGUISHING MEDIA:
SPECIAL FIRE-FIGHTING PROCEDURES:
UNUSUAL FIRE AND EXPLOSION HAZARDS:
HAZARDOUS DECOMPOSITION PRODUCTS:

SECTION 6: ACCIDENTAL RELEASE MEASURES

ACCIDENTAL RELEASE MEASURES:

SECTION 7: HANDLING AND STORAGE

HANDLING AND STORAGE:
OTHER PRECAUTIONS:

SECTION 8: EXPOSURE CONTROLS/PERSONAL PROTECTION

ENGINEERING CONTROLS:
VENTILATION:
RESPIRATORY PROTECTION:
EYE PROTECTION:
SKIN PROTECTION:
OTHER PROTECTIVE CLOTHING OR EQUIPMENT:
WORK HYGIENIC PRACTICES:
EXPOSURE GUIDELINES:

SECTION 9: PHYSICAL AND CHEMICAL PROPERTIES

APPEARANCE:
ODOR:
PHYSICAL STATE:
pH AS SUPPLIED:
pH (Other):
BOILING POINT:
F:
C:
MELTING POINT:
F:
C:
FREEZING POINT:
F:
C:

VAPOR PRESSURE (mmHg):
@
F:
C:
VAPOR DENSITY (AIR = 1):
@
F:
C:
SPECIFIC GRAVITY (H_2O = 1):
@
F:
C:
EVAPORATION RATE:
BASIS (= 1):
SOLUBILITY IN WATER:
PERCENT SOLIDS BY WEIGHT:
PERCENT VOLATILE:
BY WT/ BY VOL @
F:
C:
VOLATILE ORGANIC COMPOUNDS (VOC):
WITH WATER: LBS/GAL
WITHOUT WATER: LBS/GAL
MOLECULAR WEIGHT:
VISCOSITY:
@
F:
C:

SECTION 10: STABILITY AND REACTIVITY

STABLE UNSTABLE
STABILITY:
CONDITIONS TO AVOID (STABILITY):
INCOMPATIBILITY (MATERIAL TO AVOID):
HAZARDOUS DECOMPOSITION OR BY-PRODUCTS:
HAZARDOUS POLYMERIZATION:
CONDITIONS TO AVOID (POLYMERIZATION):

SECTION 11: TOXICOLOGICAL INFORMATION

TOXICOLOGICAL INFORMATION:

SECTION 12: ECOLOGICAL INFORMATION

ECOLOGICAL INFORMATION:

SECTION 13: DISPOSAL CONSIDERATIONS

WASTE DISPOSAL METHOD:
RCRA HAZARD CLASS:

SECTION 14: TRANSPORT INFORMATION

U.S. DEPARTMENT OF TRANSPORTATION
PROPER SHIPPING NAME:
HAZARD CLASS:
ID NUMBER:
PACKING GROUP:
LABEL STATEMENTS:
WATER TRANSPORTATION
PROPER SHIPPING NAME:
HAZARD CLASS:
ID NUMBER:
PACKING GROUP:
LABEL STATEMENTS:
AIR TRANSPORTATION
PROPER SHIPPING NAME:
HAZARD CLASS:
ID NUMBER:
PACKING GROUP:
LABEL STATEMENTS:
OTHER AGENCIES:
SECTION 14 NOTES:

SECTION 15: REGULATORY INFORMATION

U.S. FEDERAL REGULATIONS
TSCA (TOXIC SUBSTANCE CONTROL ACT):
CERCLA (COMPREHENSIVE RESPONSE, COMPENSATION, AND LIABILITY ACT):
SARA TITLE III (SUPERFUND AMENDMENTS AND REAUTHORIZATION ACT):
311/312 HAZARD CATEGORIES:
313 REPORTABLE INGREDIENTS:
STATE REGULATIONS:
INTERNATIONAL REGULATIONS:
SECTION 15 NOTES:

SECTION 16: OTHER INFORMATION

3.9.1 American Conference of Governmental Industrial Hygienists (ACGIH)

The long-time leaders in defining industrial or workplace exposure to hazardous materials is the American Conference of Government Industrial Hygienists, the ACGIH. The value and applicability of their work is recognized by the ANSI Z400 MSDS standard and the federal requirements for information that must be included on an MSDS.

Their original work introduced three new terms to HAZMAT professionals: Threshold Limit Value (TLV), Short Term Exposure Limit (STEL) and Time Weighted Average (TWA). The science and the framework have evolved considerably, and in 1988 the ACGIH formally adopted a new approach that characterized the basic concepts as TLVs which we have already identified and Biological Exposure Indices or BEI adding a new framework that characterizes TLVs in three ways: The TLV STEL; the TLV TWA; and the TLV C. While the ACGIH continues to emphasize that these are not meant to be standards, and must be interpreted and applied by qualified industrial hygienists, nonetheless the material has been incorporated into many other consensus and standards and is recognized by statutory standards and publications. A much fuller discussion on the topic can be found in Appendix A on the accompanying CD.

The author wishes to acknowledge the ACGIH for not only giving permission to use their material but for providing an updated and extremely comprehensive introduction/discussion on the topic and for their very fast and positive response for that permission.

3.9.2 National Institute for Occupational Safety and Health (NIOSH)

3.9.2.1 Introduction

The *NIOSH Pocket Guide to Chemical Hazards* is intended as a source of general industrial hygiene information for workers, employers, and occupational health professionals. The *Pocket Guide* presents key information and data in abbreviated tabular form for 677 chemicals or substance groupings (e.g., manganese compounds, tellurium compounds, inorganic tin compounds, etc.) that are found in the work environment. The industrial hygiene information found in the *Pocket Guide* should help users recognize and control occupational chemical hazards. The chemicals or substances contained in this revision include all substances for which the National Institute for Occupational Safety and Health (NIOSH) has recommended exposure limits (RELs) and those with permissible exposure limits (PELs) as found in the Occupational Safety and Health Administration (OSHA) General Industry Air Contaminants Standard (29 CFR 1910.1000).

3.9.2.2 Background

In 1974, NIOSH (which is responsible for recommending health and safety standards) joined OSHA (whose jurisdictions include promulgation and enforcement activities) in developing a series of occupational health standards for substances with existing PELs. This joint effort was labeled the Standards Completion Program and involved the cooperative efforts of several contractors and personnel from various divisions within NIOSH and OSHA. The Standards Completion Program developed 380 substance-specific draft standards with supporting documentation that contained technical information and recommendations needed for the promulgation of new occupational health regulations. The *Pocket Guide* was developed to make the technical information in those draft standards more conveniently available to workers, employers, and occupational health professionals. The *Pocket Guide* is updated periodically to reflect new data regarding the toxicity of various substances and any changes in exposure standards or recommendations.

3.9.2.2.1 Data Collection and Application

The data were collected from a variety of sources, including NIOSH policy documents such as criteria documents and Current Intelligence Bulletins (CIBs), and recognized references in the fields of industrial hygiene, occupational medicine, toxicology, and analytical chemistry.

3.9.2.3 NIOSH Recommendations

Acting under the authority of the Occupational Safety and Health Act of 1970 (29 USC Chapter 15) and the Federal Mine Safety and Health Act of 1977 (30 USC Chapter 22), NIOSH develops and periodically revises recommended exposure limits (RELs) for hazardous substances or conditions in the workplace. NIOSH also recommends appropriate preventive measures to reduce or eliminate the adverse health and safety effects of these hazards. To formulate these recommendations, NIOSH evaluates all known and available medical, biological, engineering, chemical, trade, and other information relevant to the hazard. These recommendations are then published and transmitted to OSHA and the Mine Safety and Health Administration (MSHA) for use in promulgating legal standards.

NIOSH recommendations are published in a variety of documents. Criteria documents recommend workplace exposure limits and appropriate preventive measures to reduce or eliminate adverse health effects and accidental injuries.

Current Intelligence Bulletins (CIBs) are issued to disseminate new scientific information about occupational hazards. A CIB may draw attention to a formerly unrecognized hazard, report new data on a known hazard, or present information on hazard control.

Alerts, Special Hazard Reviews, Occupational Hazard Assessments, and Technical Guidelines support and complement the other standards development activities of the Institute. Their purpose is to assess the safety and health problems associated with a given agent or hazard (e.g., the potential for injury or for carcinogenic, mutagenic, or teratogenic effects) and to recommend appropriate control and surveillance methods. Although these documents are not intended to supplant the more comprehensive criteria documents, they are prepared to assist OSHA and MSHA in the formulation of regulations.

In addition to these publications, NIOSH periodically presents testimony before various Congressional committees and at OSHA and MSHA rule-making hearings.

Recommendations made through 1992 are available in a single compendium entitled NIOSH Recommendations for Occupational Safety and Health: Compendium of Policy Documents and Statements [DHHS (NIOSH) Publication No. 92-100]. Copies of the Compendium may be ordered from the NIOSH Publications office (800-232-6348).

3.9.2.4 How to Use This Pocket Guide

The *Pocket Guide* has been designed to provide chemical-specific data to supplement general industrial hygiene knowledge. To maximize the amount of data provided in this limited space, abbreviations and codes have been used extensively. These abbreviations and codes, which have been designed to permit rapid comprehension by the regular user, are discussed for each column in the following subsections.

The chemical name found in the OSHA General Industry Air Contaminants Standard (29 CFR 1910.1000) is listed in the top left portion of each chemical table.

3.9.2.4.1 Chemical Name

The chemical name found in the OSHA General Industry Air Contaminants Standard (29 CFR 1910.1000) is listed in the top left portion of each chemical table.

3.9.2.4.2 Structure/Formula

The chemical structure or formula is listed under the chemical name in each chemical table. Carbon–carbon double bonds (–C=C–) have been indicated where applicable.

3.9.2.5 Chemical Abstracts Service (CAS) Number

This Chemical Abstracts Service (CAS) registry number is provided in the top right portion of the chemical tables. The CAS number, in the format xxx-xx-x, is unique for each chemical and allows efficient searching on computerized data bases. The CAS number index can be used to find a chemical based on the CAS number.

3.9.2.6 Registry of Toxic Effects of Chemical Substances (RTECS) Number

This section lists the Registry of Toxic Effects of Chemical Substances (RTECS®) number, in the format ABxxxxxxx. RTECS may be useful for obtaining additional toxicologic information on a specific substance. The RTECS number index can be used to find a chemical based on the RTECS number.

RTECS is a compendium of data extracted from the open scientific literature. On December 18, 2001, CDC's Technology Transfer Office, on behalf of NIOSH, successfully completed negotiating a "PHS Trademark Licensing Agreement" for RTECS. This nonexclusive licensing agreement provides for the transfer and continued development of the "RTECS Database and its Trademark" to MDL Information Systems, Inc. (MDL), a wholly owned subsidiary of Elsevier Science, Inc. Under this agreement, MDL took on the responsibility for updating, licensing, and marketing and distributing RTECS. This agreement was assigned to Symyx Technologies, Inc., in 2007, when MDL became part of Symyx. For more information visit the Symyx website.

3.9.2.7 Department of Transportation (DOT) ID and Guide Numbers

This section lists the U.S. Department of Transportation (DOT) identification numbers and the corresponding guide numbers. Their format is xxxx yyy. The identification (ID) number (xxxx) indicates that the chemical is regulated by DOT. The guide number (yyy) refers to actions to be taken to stabilize an emergency situation; this information can be found in the 2008 Emergency Response Guidebook (Pipeline and Hazardous Materials Safety Administration, U.S. Department of Transportation, East Building, 2nd Floor, 1200 New Jersey Avenue SE, Washington, DC 20590). Please note however, that many DOT numbers are *not* unique for a specific substance.

3.9.2.8 Synonyms and Trade Names

This section of each chemical table contains an alphabetical list of common synonyms and trade names for each chemical. The Chemical Name, Synonym and Trade Name Index can be used to search for chemical pages. This index also includes the primary chemical names for all of the chemicals in the *Pocket Guide*.

3.9.2.9 Conversion Factors

This section lists factors for the conversion of ppm (parts of vapor or gas per million parts of contaminated air by volume) to mg/m^3 (milligrams of vapor or gas per cubic meter of contaminated air) at 25°C and 1 atmosphere for chemicals with exposure limits expressed in ppm.

3.9.2.10 Exposure Limits

The NIOSH recommended exposure limits (RELs) are listed first in this section. For NIOSH RELs, *TWA* indicates a time-weighted average concentration for up to a 10-hour workday during a 40-hour workweek. A short-term exposure limit (STEL) is designated by "ST" preceding the value; unless noted otherwise, the STEL is a 15-minute TWA exposure that should not be exceeded at any time during a workday. A ceiling REL is designated by "C" preceding the value; unless noted otherwise, the ceiling value should not be exceeded at any time. Any substance that NIOSH considers to be a potential occupational carcinogen is designated by the notation "Ca" (see Appendix A, which contains a brief discussion of potential occupational carcinogens).

The OSHA permissible exposure limits (PELs), as found in Tables Z-1, Z-2, and Z-3 of the OSHA General Industry Air Contaminants Standard (29 CFR 1910.1000), that were effective on July 1, 1993* and which are currently enforced by OSHA are listed next.

*In July 1992, the 11th Circuit Court of Appeals in its decision in AFL-CIO v. OSHA, 965 F.2d 962 (11th Cir., 1992) vacated more protective PELs set by OSHA in 1989 for 212 substances, moving them back to PELs established in 1971. The appeals court also vacated new PELs for 164 substances that were not previously regulated. Enforcement of the court decision began on June 30, 1993. Although OSHA is currently enforcing exposure limits in Tables Z-1, Z-2, and Z-3 of 29 CFR 1910.1000 which were in effect before 1989, violations of the "general duty clause" as contained in Section 5(a)(1) of the Occupational Safety and Health Act may be considered when worker exposures exceed the 1989 PELs for the 164 substances that were not previously regulated. The substances for which OSHA PELs were vacated on June 30, 1993 are indicated by the symbol "†" following OSHA in this section and previous values (the PELs that were vacated) are listed in Appendix G.

TWA concentrations for OSHA PELs must not be exceeded during any 8-hour workshift of a 40-hour workweek. A STEL is designated by "ST" preceding the value and is measured over a 15-minute period unless noted otherwise. OSHA ceiling concentrations (designated by "C" preceding the value) must not be exceeded during any part of the workday; if instantaneous monitoring is not feasible, the ceiling must be assessed as a 15-minute TWA exposure. In addition, there are a number of substances from Table Z-2 (e.g., beryllium, ethylene dibromide) that have PEL ceiling values that must not be exceeded except for specified excursions. For example, a "5-minute maximum peak in any 2 hours" means that a 5-minute exposure above the ceiling value, but never above the maximum peak, is allowed in any 2 hours during an 8-hour workday. Appendix B contains a brief discussion of substances regulated as carcinogens by OSHA.

Concentrations are given in ppm, mg/m^3, mppcf (millions of particles per cubic foot of air as determined from counting an impinger sample), or fibers/cm^3 (fibers per cubic centimeter). The "[skin]" designation indicates the potential for dermal absorption; skin exposure should be prevented as necessary through the use of good work practices, gloves, coveralls, goggles, and other appropriate equipment. The "(total)" designation indicates that the REL or PEL listed is for "total particulate" versus the "(resp)" designation which refers to the "respirable fraction" of the airborne particulate.

Appendix C contains more detailed discussions of the specific exposure limits for certain low-molecular-weight aldehydes, asbestos, various dyes (benzidine-, *o*-tolidine-, and *o*-dianisidine-based), carbon black, chloroethanes, the various chromium compounds (chromic acid and chromates, chromium(II) and chromium(III) compounds, and chromium metal), coal tar pitch volatiles, coke oven emissions, cotton dust, lead, mineral dusts, NIAX® Catalyst ESN, trichloroethylene, and tungsten carbide (cemented). Appendix D contains a brief discussion of substances included in the *Pocket Guide* with no established RELs at this time. Appendix F contains miscellaneous notes regarding the OSHA PEL for benzene and the IDLHs for four chloronaphthalene compounds, and Appendix G lists the OSHA PELS that were vacated on June 30, 1993.

3.9.2.11 Immediately Dangerous to Life and Health (IDLH)

This section lists the immediately dangerous to life or health concentrations (IDLHs). For the June 1994 Edition of the *Pocket Guide*, NIOSH reviewed and in many cases revised the IDLH values. The criteria utilized to determine the adequacy of the original IDLH values were a combination of those used during the Standards Completion Program and a newer methodology developed by NIOSH. These criteria formed a tiered approach, preferentially using acute human toxicity data, followed by acute animal inhalation toxicity data, and then by acute animal oral toxicity data to determine a preliminary updated IDLH value. When relevant acute toxicity data were insufficient or unavailable, NIOSH also considered using chronic toxicity data or an analogy to a chemically similar substance. NIOSH then compared these preliminary values with the following criteria to determine the updated IDLH value: 10% of lower explosive limit (LEL); acute animal respiratory irritation data (RD50); other short-term exposure guidelines; and the NIOSH Respirator Selection Logic (DHHS [NIOSH] Publication No. 2005-100). The Documentation for Immediately Dangerous to Life or Health Concentrations (NTIS Publication Number PB-94-195047) further describes these criteria and provides information sources for both the original and revised IDLH values. NIOSH currently is assessing the various uses of IDLHs, whether the criteria used to derive the IDLH values are valid, and if other information or criteria should be utilized.

The purpose for establishing an IDLH value in the Standards Completion Program was to determine the airborne concentration from which a worker could escape without injury or irreversible health effects from an IDLH exposure in the event of the failure of respiratory protection equipment. The IDLH was considered a maximum concentration above which only a highly reliable breathing apparatus providing maximum worker protection should be permitted. In determining IDLH values, NIOSH considered the ability of a worker to escape without loss of life or irreversible health effects along with certain transient effects, such as severe eye or respiratory irritation, disorientation, and incapacitation, which could prevent escape. As a safety margin, IDLH values are based on effects that might occur as a consequence of a 30-minute exposure. However, the 30-minute period was NOT meant to imply that workers should stay in the work environment any longer than necessary; in fact, *every effort should be made to exit immediately!*

NIOSH Respirator Selection Logic defines IDLH exposure conditions as "conditions that pose an immediate threat to life or health, or conditions that pose an immediate threat of severe exposure to contaminants, such as radioactive materials, which are likely to have adverse cumulative or delayed effects on health." The purpose of establishing an IDLH exposure concentration is to ensure that the worker can escape from a given contaminated environment in the event of failure of the respiratory protection equipment. The NIOSH Respirator Selection Logic uses IDLH values as one of several respirator selection criteria. Under the NIOSH Respirator Selection Logic, the most protective respirators (e.g., a self-contained breathing apparatus equipped with a full facepiece and operated in a pressure-demand or other positive-pressure mode) would be selected for firefighting, exposure to carcinogens, entry into oxygen-deficient atmospheres, in emergency situations, during entry into an atmosphere that contains a substance at a concentration greater than 2000 times the NIOSH REL or OSHA PEL, and for entry into IDLH atmospheres. IDLH values are listed in the *Pocket Guide* for over 380 substances.

The notation "Ca" appears in the IDLH field for all substances that NIOSH considers potential occupational carcinogens. However, IDLH values that were originally determined in the Standards Completion Program or were subsequently revised are shown in brackets following the

"Ca" designations. "10%LEL" indicates that the IDLH was based on 10% of the lower explosive limit for safety considerations even though the relevant toxicological data indicated that irreversible health effects or impairment of escape existed only at higher concentrations. "N.D." indicates that an IDLH value has not been determined for that substance. Appendix F contains an explanation of the "Effective" IDLHs used for four chloronaphthalene compounds.

3.9.2.12 Physical Description

This entry provides a brief description of the appearance and odor of each substance. Notations are made as to whether a substance can be shipped as a liquefied compressed gas or whether it has major use as a pesticide.

3.9.2.13 Chemical and Physical Properties

The following abbreviations are used for the chemical and physical properties given for each substance. "NA" indicates that a property is not applicable, and a question mark (?) indicates that it is unknown.

MW	Molecular weight
BP	Boiling point at 1 atmosphere, °F
Sol	Solubility in water at 68°F (unless a different temperature is noted), % by weight (i.e., g/100 ml)
Fl.P	Flash point (i.e., the temperature at which the liquid phase gives off enough vapor to flash when exposed to an external ignition source), closed cup (unless annotated "(oc)" for open cup), °F
IP	Ionization potential, eV (electron volts) [Ionization potentials are given as a guideline for the selection of photo ionization detector lamps used in some direct-reading instruments.]
VP	Vapor pressure at 68°F (unless a different temperature is noted), mmHg; "approx" indicates approximately
MLT	Melting point for solids, °F
FRZ	Freezing point for liquids and gases, °F
UEL	Upper explosive (flammable) limit in air, % by volume (at room temperature unless otherwise noted)
LEL	Lower explosive (flammable) limit in air, % by volume (at room temperature unless otherwise noted)
MEC	Minimum explosive concentration, g/m^3 (when available)
Sp.Gr	Specific gravity at 68°F (unless a different temperature is noted) referenced to water at 39.2°F (4°C)
RGasD	Relative density of gases referenced to air = 1 (indicates how many times a gas is heavier than air at the same temperature)

When available, the flammability/combustibility of a substance is listed at the bottom of the chemical and physical properties section. The following OSHA criteria (29 CFR 1910.106) were used to classify flammable or combustible liquids:

Class IA flammable liquid	Fl.P. below 73°F and BP below 100°F.
Class IB flammable liquid	Fl.P. below 73°F and BP at or above 100°F.
Class IC flammable liquid	Fl.P. at or above 73°F and below 100°F.
Class II combustible liquid	Fl.P. at or above 100°F and below 140°F.
Class IIIA combustible liquid	Fl.P. at or above 140°F and below 200°F.
Class IIIB combustible liquid	Fl.P. at or above 200°F.

3.9.2.14 Incompatibilities and Reactivities

This entry lists important hazardous incompatibilities or reactivities for each substance.

3.9.2.15 Measurement Methods

The section provides a source (NIOSH or OSHA) and the corresponding method number for measurement methods which can be used to determine the exposure for the chemical or substance. Unless otherwise noted, the NIOSH methods are from the 4th edition of the NIOSH Manual of Analytical Methods (DHHS [NIOSH] Publication No. 94-113 and supplements). If a different edition of the NIOSH Manual of Analytical Methods is cited, the appropriate edition and, where applicable, the volume number are noted (e.g., II-4 [2nd edition, volume 4]). The OSHA methods are from the OSHA Web site ("http://www.osha.gov/dts/sltc/methods/index.html"). "None available" means that no method is available from NIOSH or OSHA.

Each method listed is the recommended method for the analysis of the compound of interest. However, the method may not have been fully optimized to meet the specific sampling situation. Note that some methods are only partially evaluated and have been used in very limited sampling situations. Review the details of the method and consult with the laboratory performing the analysis regarding the applicability of the method and the need for further modifications to the method in order to adjust for the particular conditions.

3.9.2.16 Personal Protection and Sanitation Recommendations

This section presents a summary of recommended practices for each substance. These recommendations supplement general work practices (e.g., no eating, drinking, or smoking where chemicals are used) and should be followed if additional controls are needed after using all feasible processes, equipment, and task controls. Each category is described as follows:

SKIN:	Recommends the need for personal protective clothing.
EYES:	Recommends the need for eye protection.
WASH SKIN:	Recommends when workers should wash the spilled chemical from the body in addition to normal washing (e.g., before eating).
REMOVE:	Advises workers when to remove clothing that has accidentally become wet or significantly contaminated.
CHANGE:	Recommends whether the routine changing of clothing is needed.
PROVIDE:	Recommends the need for eyewash fountains and/or quick drench facilities.

3.9.2.17 First Aid

This entry lists emergency procedures for eye and skin contact, inhalation, and ingestion of the toxic substance.

3.9.2.18 Respirator Selection Recommendations

This section provides a condensed table of allowable respirators to be used for those substances for which IDLH values have been determined, or for which NIOSH has previously provided respirator recommendations (e.g., in criteria documents or Current Intelligence Bulletins) for certain chemicals. There are, however, 186 chemicals listed in the *Pocket Guide* for which IDLH values have yet to be determined. Since the IDLH value is a critical component for completing the NIOSH Respirator Selection Logic for a given chemical, the *Pocket Guide* does not provide respiratory recommendations for those 186 chemicals without IDLH values. As new or revised IDLH values are developed for those and other chemicals, NIOSH will provide appropriate respirator recommendations. (Appendix F contains an explanation of the "Effective" IDLHs used for four chloronaphthalene compounds.)

In 1995, NIOSH developed a new set of regulations in 42 CFR 84 [PDF] (also referred to as "Part 84") for testing and certifying nonpowered, air-purifying, particulate-filter respirators. The new Part 84 respirators have passed a more demanding certification test than the old respirators (e.g., dust; dust and mist; dust, mist, and fume; spray paint; pesticide) certified under 30 CFR 11 (also referred to as "Part 11"). Recommendations for nonpowered, air-purifying particulate respirators have been updated from previous editions of the *Pocket Guide* to incorporate Part 84 respirators; Part 11 terminology has been removed.

In January 1998, OSHA revised its respiratory protection standard (29 CFR 1910.134). Among the provisions in the revised standard is the requirement for an end-of-service-life indicator (ESLI) or a change schedule when air-purifying respirators with chemical cartridges or canisters are used for protection against gases and vapors [29 CFR 1910.134(d)(3)(iii)] requirement. In the *Pocket Guide*, air-purifying respirators (without ESLIs) for protection against gases and vapors are recommended only for chemicals with adequate warning properties, but now these respirators may be selected regardless of the warning properties. Respirator recommendations in the *Pocket Guide* have not been revised in this edition to reflect the OSHA requirements for ESLIs or change schedules.

Appendix A lists the NIOSH carcinogen policy. Respirator recommendations for carcinogens in the *Pocket Guide* have not been revised to reflect this policy; these recommendations will be revised in future editions.

The first line in the entry indicates whether the "NIOSH" or the "OSHA" exposure limit is used on which to base the respirator recommendations. The more protective limit between the NIOSH REL or the OSHA PEL is always used. "NIOSH/OSHA" indicates that the limits are equivalent.

Each subsequent line lists a maximum use concentration (MUC) followed by the classes of respirators, with their Assigned Protection Factors (APFs), that are acceptable for use up to the MUC. Individual respirator classes are separated by diagonal lines (/). More protective respirators may be worn. "Emergency or planned entry into unknown concentrations or entry into IDLH conditions" is followed by the classes of respirators acceptable for these conditions. "Escape" indicates that the respirators are to be used only for escape purposes. For each MUC or condition, this entry lists only those respirators with the required APF and other use restrictions based on the NIOSH Respirator Selection Logic.

In certain cases, the recommended respirators are annotated with the following symbols as additional information:

* Substance reported to cause eye irritation or damage; may require eye protection
£ Substance causes eye irritation or damage; eye protection needed
¿ Only nonoxidizable sorbents allowed (not charcoal)
† End of service life indicator (ESLI) required

All respirators selected must be approved by NIOSH under the provisions of 42 CFR 84. The current listing of NIOSH/MSHA certified respirators can be found in the NIOSH Certified Equipment List.

A complete respiratory protection program must be implemented and must fulfill all requirements of 29 CFR 1910.134 respiratory protection program must include a written standard operating procedure covering regular training, fit-testing, fit-checking, periodic environmental monitoring, maintenance, medical monitoring, inspection, cleaning, storage, and periodic program evaluation. Selection of a specific respirator within a given class of recommended respirators depends on the particular situation; this choice should be made only by a knowledgeable person. *REMEMBER:* Air-purifying respirators will not protect users against oxygen-deficient atmospheres, and they are *not to be used in IDLH conditions.* The only respirators recommended for firefighting are self-contained breathing apparatuses that have full facepieces and are operated in a pressure-demand or other positive-pressure mode. Additional information on the selection and use of respirators can be found in the NIOSH Respirator Selection Logic (DHHS [NIOSH] Publication No. 2005-100) and the NIOSH Guide to Industrial Respiratory Protection (DHHS [NIOSH] Publication No. 87-116).

3.9.2.19 Exposure Route, Symptoms, Target Organs

3.9.2.19.1 Exposure Route

This section lists the toxicologically important routes of entry for each substance and whether contact with the skin or eyes is potentially hazardous.

3.9.2.19.2 Symptoms

This entry lists the potential symptoms of exposure and whether NIOSH considers the substance a potential occupational carcinogen.

3.9.2.19.3 Target Organs

This entry lists the organs that are affected by exposure to each substance. For carcinogens, the type(s) of cancer are listed in brackets. Information in this section reflects human data unless otherwise noted.

3.9.2.20 Selection of N-, R-, or P-Series Particulate Respirators

1. The selection of N-, R-, and P-series filters depends on the presence of oil particles as follows:
   - If no oil particles are present in the work environment, use a filter of any series (i.e., N-, R-, or P-series).

- If oil particles (e.g., lubricants, cutting fluids, glycerin) are present, use an R- or P-series filter. *Note: N-series filters cannot be used if oil particles are present.*
- If oil particles are present and the filter is to be used for more than one work shift, use only a P-series filter.

Note: To help you remember the filter series, use the following guide:

N for **N**ot resistant to oil,
R for **R**esistant to oil,
P for oil **P**roof.

2. Selection of filter efficiency (i.e., 95%, 99%, or 99.97%) depends on how much filter leakage can be accepted. Higher filter efficiency means lower filter leakage.
3. The choice of facepiece depends on the level of protection needed—that is, the assigned protection factor (APF) needed.

3.10 Emergency Response Guide (ERG)

The ERG (see Table 3.2) was first issued by the US Department of Transportation in 1973 by the then Research and Special Programs Administration now known as *PHMSA, Pipeline and Hazardous Materials Safety Administration* it has been reissued every four years since 1992, i.e., 1996, 2000, 2004, 2008, and 2012.

It started out an American publication; by 2000 it was American, Canadian, and Mexican, and in 2008 it was American, Canadian, Mexican, Argentinean, Brazilian, and Columbian.

It has been translated into more languages than just about everything except the Bible.

In 2004 1.73 million ERGs were printed by the US DOT, 2.2 million ERGs were planned for the 2008 version. It is distributed free "for every emergency vehicle in the U.S." but they typically do not reprint, so you may buy them from the GPO and commercial sources.

The ERG is based on the internationally recognized hazard classification and identification system.

CLASSIFICATION
Nine Hazard classes with a total of 25 classes and divisions

IDENTIFICATION
Six-digit letter–number combination
Labels and placards with symbols, colors, and numbers
Shipping papers with three elements of interest: proper shipping name (PSN), ID number, and class

Table 3.2 Organization of the Emergency Response Guide (ERG)

The ERG is organized into five (5) sections:	
White Pages →	How to use the guide/General Information Pipeline Terrorism explosives information
Yellow Pages →	Numerical Listing by ID number
Blue Pages →	Alphabetical Listing by Proper Shipping Name (PSN)
Orange Pages →	Guide # with EMERGENCY RESPONSE INFORMATION
Green Pages →	Initial isolation table for Inhalation Hazards

There are millions of different hazardous materials. Approximately 4000 ID numbers are currently used in the various regulations, and there have been only 61 guide numbers in the ERG since 1973. However, in 2008 a new guide number was added for lithium-ion batteries; at the current time, there are 62 Guides in the ERG.

THEREFORE—

You need to read the ERG very carefully before engaging in any action.

The ERG is good for the "first minute."

The ERG four-color coded sections—yellow, blue, orange, and green—help in identifying hazardous material and determining what action to take.

In addition, there is a great deal of helpful information in the White Pages at the beginning, as well as more information and a glossary in the White section at the back of the guide.

The White front section explains how to use the guide and gives pictorial examples of standard rail and highway tanks used to transport specific types of HAZMAT, along with some dangers associated with certain types of HAZMAT.

The rear White section covers terrorist activity recognition and a glossary of terms.

TO USE THE ERG:

You need the PSN or the last four digits of the ID number.

If you have the PSN, you can go to the Blue section to get a guide number.

If you have the 4-digit number, you can go to the Yellow section to get the guide number.

In either case you need to note if the item is highlighted. If an item is highlighted and there is no fire, you MUST go to the Green section as you have something that presents a definite inhalation risk.

If an item is NOT highlighted, or after you have followed the green section you need to stop and read the appropriate orange guide in its entirety—both pages.

In the United States you may contact the PHMSA regional specialist to provide formal presentations on the use of the Emergency Response Guide; the 2011 program consists of a 53-slide PowerPoint presentation.

3.10.1 How to Use the Emergency Response Guide

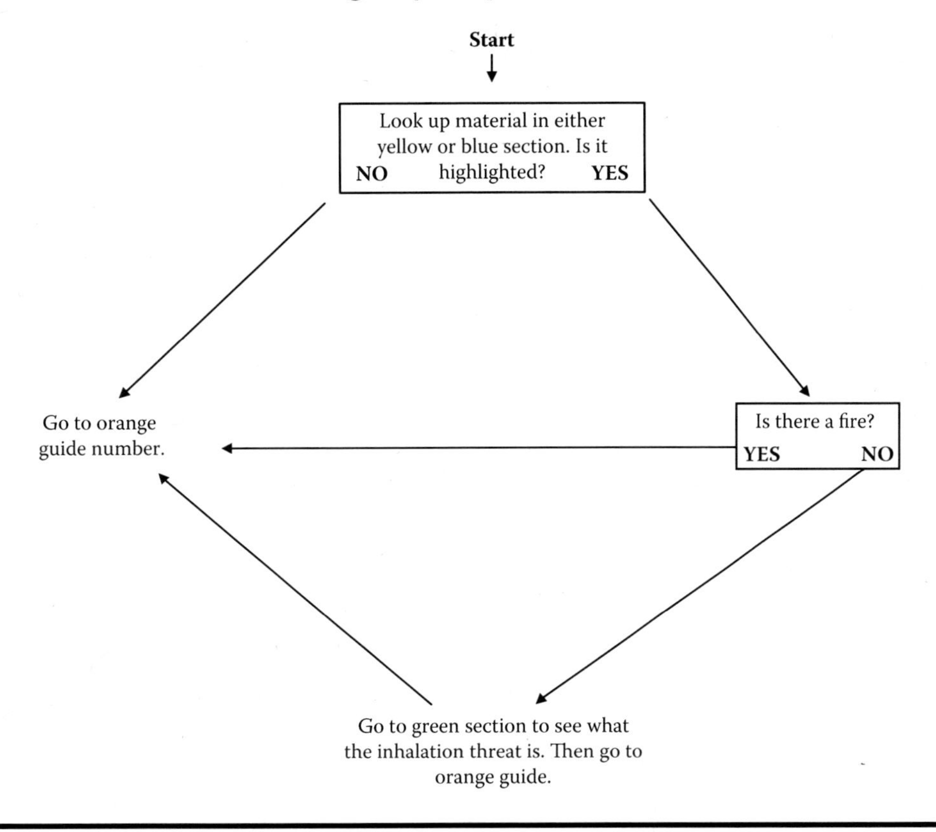

Figure 3.3 A step-by-step guide to using the ERG.

References

Bierma, T. & Waterstraat, F. (2000). *Chemical management.* New York: John Wiley & Sons.
Griffin, R. D. (2009). *Principles of hazardous materials management.* Boca Raton, FL: Taylor & Francis.
IATA. (2012). *Dangerous goods regulations.* Montreal, CA: IATA.
ICAO. (2010). *Technical instructions for the safe transportation of dangerous goods by air.* Montreal: ICAO.
IMO. (2010). *International maritime dangerous goods code.* London: IMO.
UN. (2009). *Recommendations on the transport of dangerous goods.* New York: UN.
UN. (2011). *Globally harmonized system of classification and labelling of chemicals (GHS).* New York: UN.
U.S. EPA. (2012 February 6). *Brownfields and Land Revitalization.* Retrieved from http://epa.gov/brownfields/.

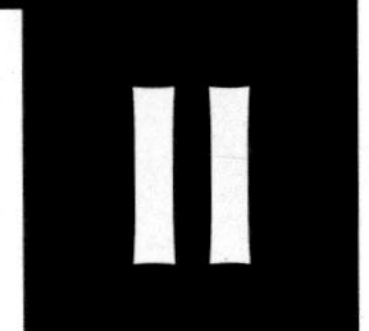

DEVELOPING PROGRAMS FOR BROAD SECTORS

Important note:

Chapter 4, Chapter 5, and Chapter 6 are designed to act as complete standalone units for specific sectors. As a result, common material is repeated in each chapter.

Chapter 4

Private Business Issues, Challenges, Opportunities, and Solutions

4.1 Initial Framework for Further Discussion

Figure 4.1 represents the basic model we will use for this entire part; the names of the organizational components may change slightly and the number may even be altered, but the basic conceptual framework will remain the same: an organization is made up of identifiable, if arbitrary, major components. In this particular chapter there are four identifiable blocks that interlock to make the organization. As we look at other areas, the titles for the blocks and the number of blocks might change, but conceptually each organization can be dissected in a similar analogous manner not necessarily a direct duplication of the areas labeled above. The object here is to divide the organization into units or areas that can be examined with respect to hazardous materials, and, the large area of waste materials and waste streams. It is now recognized that streams traditionally considered simply as "waste" can be looked at as having a negative impact on the environment and in some cases generating directly, or indirectly, hazardous conditions in the environment. For that reason we will go beyond what would traditionally be defined as hazardous materials under any regulation or concept but look at the large impact of wastes on the planetary ecosystem.

Some of the most current thinking on "business" recognizes two key concepts that Figure 4.2 tries to capture. First is the concept that all human endeavors are based upon "processes." The second is that, even in the smallest of organizations, the output can also be described as a result of a system of systems. A third concept, which I will not even try to define, is the concept of "green." Whatever the concept, construct, or paradigm (pick your choice), they will underline and connect much of the material here. The first two concepts led to development of Figure 4.2. Figure 4.2 is a supporting and an alternative approach to the concept conveyed in Figure 4.1 so it is important to discuss both when one examines hazardous materials, environmental impact, and life cycle management. We hear the expressions "going green," "green technologies," "green jobs and technologies," and "green communities and green buildings." At this point in time, there are no universally

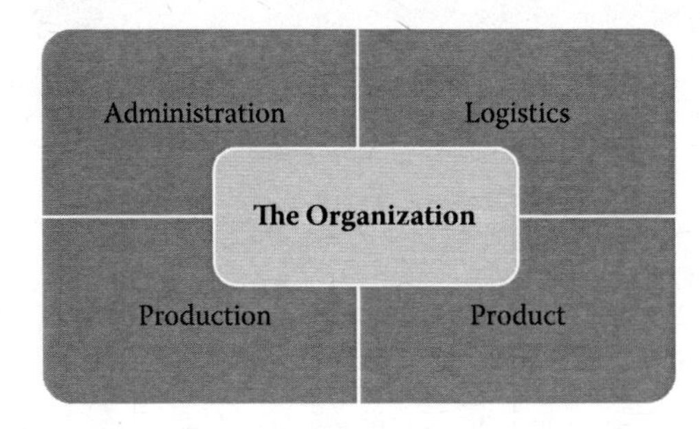

Figure 4.1 Functional view of a business enterprise.

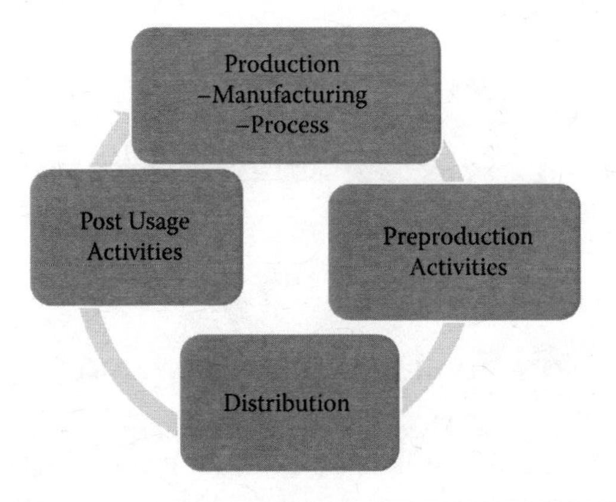

Figure 4.2 A flow diagram for business activities.

agreed-upon definitions or even a reasonably small number of definitions to cover each of those terms, but conceptually, in addition to lowering energy use, they all are addressing the underlying basis for life cycle management and the concept of hazardous materials life cycle management. This text uses the broadest possible interpretation of "hazardous materials" here.

Any system designed to segregate areas is by definition arbitrary. For the purposes of this chapter, the primary categories we will use are administration, logistics, production, and product. These in turn will be made up of a mixture of functions that include:

1. Office functions
2. Maintenance and janitorial functions
3. Product-related functions
4. Process-related functions
5. Manufacturing-related functions, both heavy and light
6. Facility-related functions

Each of the four broad "functional areas" depicted in Figure 4.1 are made up or consist of a mixture of the "functions" numbered one through six.

Let's take a few practical examples to help understand the application of Figure 4.1, the functional areas and the component functions.

Table 4.1 Functional Areas and Component Functions of an Automobile Manufacturing Facility

Function	*Example areas*	*Functional components*
Administration	Office work including: HR Engineering IT Finance Product development and design Product engineering Facility engineering	Office functions Maintenance and janitorial functions Facility-related functions
Logistics	Product development and design Purchasing Scheduling Shipping Receiving	Office functions Maintenance and janitorial functions Facility-related functions
Production	Machining Assembly Testing QA	Maintenance and janitorial functions Product-related functions Process-related functions Manufacturing related functions both heavy and light Facility-related functions
Product	Design and engineering Component part selection Assembly processes Assembly materials	Product-related functions

4.1.1 Large Automobile Manufacturing Facility

See Table 4.1.

4.1.2 Consumer Electronics Manufacturer

See Table 4.2.

4.1.3 Retail Packaged Food Producer

See Table 4.3.

As you can see, there are really only very slight differences in these three matrices. That would seem to demonstrate that this is a reasonable framework that can be applied across a very large

Table 4.2 Functional Areas and Component Functions of Consumer Electronics Manufacturer

Function	*Example areas*	*Functional components*
Administration	Office work including: HR Engineering IT Finance Product development and design Product engineering Facility engineering	Office functions Maintenance and janitorial functions Facility-related functions
Logistics	Product development and design Purchasing Scheduling Shipping Receiving	Office functions Maintenance and janitorial functions. Facility-related functions
Production	Machining Etching Plating Assembly Testing QA	Maintenance and janitorial functions Product-related functions Process-related functions Manufacturing related functions light Facility-related functions
Product	Design and engineering Component part selection Assembly processes Assembly materials Return repair policy	Product-related functions

spectrum of commercial enterprises. With minor adjustments, this tool is just as effective in the public sector and in the NGO not-for-profit world.

Nonetheless, the choice of terms and parameters masks the fact that differences between businesses most certainly do exist. The broad terms used here hide significantly different functions, activities, actions, risks, and liabilities. Here are some examples that demonstrate vastly different issues that directly relate to hazardous materials and therefore life cycle management, between some key broad business sectors:

A. A food processor would have to invest considerably in chemical laboratory testing as well as some form of cooking process. This would almost definitely require using high heat and therefore consuming a lot of energy. Nutritional information as well as allergic reactions as well as subjective taste testing would be parts of the overall production process.

Table 4.3 Functional Areas and Component Functions of a Retail Food Producer

Function	*Example areas*	*Functional components*
Administration	Office work including: HR IT Finance Product development and design Facility engineering	Office functions Maintenance and janitorial functions Facility-related functions
Logistics	Product development and design Purchasing Scheduling Shipping Receiving	Office functions Maintenance and janitorial functions Facility-related functions
Production	Coating Preparation Processing Packaging Testing QA	Maintenance and janitorial functions Product-related functions Process-related functions Manufacturing-related functions light Facility-related functions
Product	Design and engineering Component part selection Assembly processes Assembly materials Return repair policy	Product-related functions

B. A consumer electronics manufacturer might be very involved with manufacturing printed circuit boards, which is essentially an acid etching process, and/or with doing some forms of plating or surface treatment. Plating typically involves the use of arsenical processes. Polishing would be another issue for a consumer electronics product.

C. For many processed or manufactured foods, the package may be as expensive or more expensive than the product (think small packages of table salt, toothpicks, or even condiments for the food you buy at fast food restaurants). In many of these cases, the package or packaging is the most critical aspect of the product and that packaging must fulfill multiple marketing roles. For foods requiring little or no processing, the packaging may be more expensive and more critical than the process or the product.

D. For producers of clothing, there is a completely different set of criteria that become critical. Perhaps one that is most often overlooked is the actual dying of materials to give them their color. There are many natural organic dyes but for both clothing manufacturers and food manufacturers many dyes present their own health, safety, and environmental risks. For the clothing manufacturers, the dying process is energy intensive and waste producing, while introducing considerable additional liabilities for the company and much larger environmental impacts for the company to deal with.

E. On the other hand, consumer electronics packaging has two distinct meanings, neither of which can be easily matched against the food production process. The physical object that such a company produces is the "package" but to intermediaries and final end-users, the preparation of that object into a retail packaging is the package that the consumer sees and responds to, and which often leads to the sale at the retail outlet. Yet a third level of packaging is the intermediate, or tertiary, packaging used in order to allow that retail product to move halfway around the world and appear in a retail outlet.

F. Fire retardants and fire-resistant processes and materials are areas routinely overlooked in most business discussions but they have already had, and will continue to have, a very significant impact on society. Example 1: The toxic fumes from fire-retardant material used in passenger aircraft cabins has caused more airline passenger deaths than any other single source, outside of aircraft failures at altitude, over the past two to three decades. Example 2: Fire safety and firefighting, in the days when structures and all their interior accoutrements were natural, prescribed procedures for escape included getting as close to the floor as possible to prevent inhaling overheated air that could sear the lungs and kill. While it is not true in every case, the majority of toxic fumes tend to be heavier than air so they settle toward the floor. Today, the proper procedure requires individuals to stay at or slightly below door knob height. Within the structure itself, especially residences or high-occupancy commercial structures, the inherent value of sprinklers is greatly increased for this reason alone. From a firefighter's standpoint, there are a number of negatives associated with increased use of fire-resistant and fire-retardant materials within or as part of the structure. The big advantage is that delay or elimination of open flames allows for more response time. The negatives include the generation of toxic fumes which in turn requires more personal protective equipment, thus slowing down firefighters, and, in some cases possibly preventing underprepared ill-equipped firefighting attempts. Fire-retardant and fire-resistant materials also tend to hold in heat, making the entire firefighting process more difficult and the ensuing overhaul times longer and more water intensive. At the same time increased use of these sorts of materials makes venting fires more critical. However, that very act also tends to feed more oxygen to the fire, although it clearly reduces heat, thus reducing the possibility of flashover. We are not going to get into fire science here, but the key point is that modern materials create many challenges to the firefighter. The old ways of fire fighting and the fire safety protocols for structure fires have changed greatly simply because of new materials used to build and outfit/equip structures.

4.2 Recognizing Defined and Undefined Hazardous Materials as Material Streams and Hazardous Waste Streams

Using the framework presented in Figure 4.1, it is possible to construct a matrix that identifies activities and functions that can directly impact life cycle outcomes. See Table 4.4.

The table is a tool for exploring hazardous streams in their broadest sense while still allowing us to recognize contributions from the earliest stages of product identification through development, all the way to customer/consumer use and post-sales disposal. The laws governing many of the issues relate to hazardous materials "ownership" and therefore management. Local, federal,

Table 4.4 The Organizational Construct

Administrative functions	**Exposure risk opportunity**
Intellectual property development Administrative/office Facility management and maintenance Janitorial Support operations	Management of design: understanding of practices and materials to be avoided Feedback from production service and logistics Office supply quantities and materials Lighting, heating, and cooling Utilities and space Cleaning supplies and disposal methods
Logistics functions	**Exposure risk opportunity**
Purchasing of raw materials Purchasing of parts and supplies and consumables Product packaging Transportation packaging Distribution functions including storage warehousing transportation	Product selection criteria going beyond pricing Transportation packaging Material handling equipment Transportation methods Transportation providers
Production functions	**Exposure risk opportunity**
Machining processes Joining and fastening processes Chemical processes (plating, etching, coating, or lamination for example) Lubrication and/or testing of finished products	Lubricating and cooling chemicals Fasteners, methods, and products Surface finishers such as paint or lacquer or other sealers and treatments such as passivation Chemical processes; chemical selection and chemical recovery; treatment Material usage and selection
Product functions	**Exposure risk opportunity**
Product cleaning and maintenance Product accessibility Product shelf packaging Product preservation Batteries	Product packaging Need for and choice of operation and maintenance consumables and replaceable Product packaging and preservation design and application

and international statutory requirements are changing and expanding to address, and assign responsibility for, total cradle-to-grave management of hazardous materials and hazardous emissions/waste streams. Today, manufacturing entities are slowly being forced to both factor in and identify/include costs for final disposal, so it is important to examine the entire life cycle in order to understand how early in the process such considerations must be factored into the equation.

4.2.1 Administrative Functions

4.2.1.1 Management of Design; Understanding of Practices and Materials to Be Avoided

The earliest stages of a development cycle for a product may go by many different titles, but they do include product concept, initial product considerations, and, in a very loose sense, initial product design. At this early stage, it is important to recognize the disadvantages and risks associated with many items that are often taken for granted. This would involve:

- Understanding the inherent risks of newer technology products
- Acknowledging changes in technology that are eliminating the use of materials that in the past would have been considered hazardous wastes
- Having a basic awareness and understanding of materials engineering, that is, composite materials instead of metals

Aesthetic issues—which may be major issues for consumer-based products versus the costs of reaching satisfactory aesthetic results—also need to be factored in early in the product development cycle. There is no right or wrong answer and any specific choice must be governed by many considerations, not only the hazardous materials and environmental considerations. The issue here is to understand the inherent advantages and disadvantages of materials, processes, and design standards, as well as manufacturing and production standards in order to balance out pros and cons for each individual product.

4.2.1.2 Feedback From Production, Service, and Logistics

The second area that deserves more attention than is currently given is feedback from the entire supply chain with a heavy focus on the tail end of the production chain, the logistics chain, and the user chain. A very large number of retail products sold today are the equivalent of second, third, or even later generations of similar devices. Most of these represent changes that have been driven by technology, and cost, or, changes in consumer tastes rather than specifically including the driver of negative feedback from the entire life cycle process. Such information might lead to changes in design, changes in specific components, changes in assembly techniques, changes in the method of fastening, changes in the packaging for the product, or, possibly significant change in design criteria for a product. A complex product typically requires a week of training, and simple design changes can reduce that week of training to three days. There are huge benefits for the manufacturing organization as well as the using organization. For many industrial products, especially production machinery and testing instrumentation, a design change may reduce the routine maintenance time for replacement of consumable items, such as batteries and gas, oil, or other filters/filter elements, O-rings, or other shelf life items. Maintenance times can be reduced from hours to minutes, representing huge cost-saving opportunities for the using organization. Remember that this translates into a huge competitive advantage for the manufacturing organization and all the intermediaries in the distribution chain.

Another much more subtle issue is in a change in the entire culture of product design and development. Logistics have advanced to the degree that supply chain professionals need to be given more latitude in the selection of component parts. Designers need to shift to a functional approach for identifying parts rather than the more traditional tendency to over-specify component parts.

In the past, the designer might have routinely talked about something such as a "3 AG 1/10 A fuse." In reality, today's supply chain needs to understand the function to be performed, which is to protect the circuit if more than a certain amount of current flows. In that case, the designer needs to indicate circuit protection at the 1/10 of an ampere level. There are also quick-acting, slow-acting, and normal fuses. Again, a designer needs to make clear what the need is rather than identifying the product based on familiarity. It is a supply chain specialist—in this case, the procurement specialist—who should be going in to determine, based on multiple company criteria, what the best choice might be. There may very well need to be interchange between these two professionals based on form-fit function or the design engineering team, but, in the end, component decision may or should shift to the supply chain professional rather than the engineer. There are many exceptions, and this is an oversimplification but, conceptually, it is in court point that needs to be introduced early in discussion aesthetic company culture never overlooks opportunities for efficiency and cost savings.

4.2.1.3 Office Supply Quantities and Materials

Within the larger construct of administrative functions, office-supply quantities and materials are an area that very few companies examine closely in terms of risk and vulnerability in regards to hazardous materials and environmental impact. While everyone recognizes the inherent economic advantages of bulk purchasing office supplies, very few recognize just how many are actually hazardous in nature or regulated as hazardous wastes for commercial disposal. This is an area that is far more complex than we can treat here but to give you a rough idea, this includes all of the following products: all forms of fluorescent light bulbs, any mercury light bulbs, all batteries, most toners for copying machines, "whiteout," dry-erase markers, many whiteboard and other electronic product cleaners and some, but certainly not all, adhesives used in an office setting. While many in the Western world might not understand or recognize "carbon paper," the very name should warn everyone that there is an inherent liability issue here, and in fact real carbon paper is a regulated item for that very reason and represents multiple hazards in transport and storage and for disposal. The easiest but certainly not a foolproof way to determine what risks in this area a product might represent is to simply inquire as to whether or not it has an MSDS. If an item has an MSDS, there has to be a reason. Granted, in some cases the hazard may be one your school or business does not need to worry about, but it is still your responsibility and ultimately the risk or liability does fall back on the business or school if an untoward occurrence brings a suit.

4.2.1.4 Lighting, Heating, and Cooling

There are number of separate areas here, and some of what we talked about will be repeated when we address production issues. Technology is also changing at a rapid rate, so material that may be correct at the time of publication may no longer conform in as short a period as six months to a year.

The United States as a nation and as a market is shifting from incandescent lighting to other forms of lighting. A lot of that shift has already occurred in commercial and industrial spaces so legislation-driven changes will most heavily impact consumers and private residences in the current near term. In the commercial marketplace, fluorescent lighting has been used for years. What many may be unaware of is that fluorescent lights ballasts containing PCBs and starters from earlier units represented regulated waste streams, although their inherent hazards could be totally ignored as part of the initial sales or installation process. What all should be aware of is that a "cottage industry" came into existence some time ago, once the inherent hazards associated with lights, ballasts containing BCBs, and fluorescent tubes became recognized. In the 1950s and 1960s it is unlikely anyone recognized the looming need to design and manufacture low-cost but effective packaging to return burned-out fluorescent lights. Today that is the norm; those lights need to retain their integrity until they arrive at a properly designated and licensed HAZMAT recycling facility. That means breakage must be avoided at all costs. Today's current family of CFLs (compact fluorescent lights), which are rapidly replacing incandescent light bulbs, are presenting the same hazards as fluorescence have always represented, that is primarily a mercury vapor hazard. That means there is this liability, as well as costs, attached to the use of all forms of fluorescent lighting. Assuming that one does not or cannot obtain incandescent light bulbs (ILBs), the alternatives include mercury and halogen as well as LEDs. At this point in time, remembering the caveat above and regardless of the acquisition cost, it would appear that LEDs provide the greatest opportunity to reduce liability and risk, and for eliminating hazardous material waste streams. LEDs are currently quite expensive but they have outstanding lifetime characteristics and to the best of the author's knowledge represent no form of hazardous waste stream when they reach the end of their service life. Mercury light bulbs clearly represent the same hazards as all fluorescent bulbs; please note the word "mercury" in their name. Halogen lights have enjoyed and continue to enjoy a fair bit of popularity. From the life cycle, operation cost, safety, and ecological standpoints, halogens present the least attractive alternate light source. Halogens tend to run at very high temperatures and that alone makes them less desirable because the high temperatures they generate represent a large waste of energy and a theoretically higher risk as an ignition source that could start structural fires.

Let us talk a little bit more about lighting without becoming engineers or scientists. According to one *National Geographic* article, most incandescent light sources are between 1.9% and 2.6% efficient while fluorescent light sources are typically 9% to 11% efficient, 3 to 4 times more efficient than incandescent (Moll, 2012). Current claims are that LEDs are 15 times more efficient than incandescent lights. That makes LEDs very competitive since there are no hazardous materials in the product, they run cooler, are more efficient, and therefore present less risk and liability both in operation and for disposal. In reality, incandescent light bulbs are heaters that also provide light; that is especially true when talking about halogens. What is overlooked in most discussions are the true life cycle costs for the different forms of lighting. In colder climates, where in the past incandescent lighting was providing incremental heating, a switch from incandescent to fluorescent can actually increase greenhouse gases, increase total life cycle costs and therefore have yet another detrimental effect on the environment and the economy. Assuming one could continue to use incandescent lights and the facility was in the appropriate climate, such as those with four seasons, and above or below certain altitudes, over a vast portion of the globe, or, incandescent lighting might provide the best set of trade-offs. If that is true, a lot of research and a lot of engineering has to go into the actual facility to maximize benefits and minimize costs. With sufficient opportunity to help meet a heating load, even seasonally, incandescent lights may make a much better choice than business has been led to believe, and unfortunately better than what government and environmentalists think they understand. At all sorts of facilities closer to the equator one must look at LEDs as the preferred alternative, based on science not the propaganda put out by many organizations including governments. Switching from ILBs to CFLBs may not always result in an environmentally friendly outcome, "especially in cold climates," as the authors of a Canadian report conclude (Ivanko, Waher & Warbody, 2009).

Heating is another extremely complex area; for the vast majority of businesses it can be placed lower on the list. Realistically, individual operations are going to have to make a number of different choices about heat. The primary choices are either electrical heating or some form of fossil fueled system. Based on the technologies available in 2012, choices for many smaller businesses in their own facilities are extremely broad. For those in leased or rented spaces, utilities may or may not be included in such payments and therefore choices may be more limited. For those who are planning on using leased or rented spaces, consideration of heating and cooling sources should be a factor in selecting a location and might even need to be looked at from an economic versus environmental cost trade-off standpoint. For those who are actually going to design or build their own new facilities, both geothermal and solar systems offer reasonable alternatives either as primary systems or auxiliary/backup systems. The bottom line is that from an environmental and life cycle standpoint, natural or liquefied petroleum gas offers significant advantages over all other forms. Today additional sources are being developed which might replace any or all of these, depending on the size of the operation, but all need to be considered and the positives and negatives in terms of hazards to the air, to the Earth, and to waste streams have to be examined. The larger issue is what sort of operations go on in the business and whether there is a way to recover large quantities of otherwise wasted heat to offset the need to generate additional heat in order to maintain a viable internal environment. As an example, and an historic reference point, in the days before digital electronics, it was not unusual for military communications facilities to be designed to recapture the heat from tube-operated equipment to maintain reasonable operating temperatures within their facilities. The Naval Communication Station in Adak, Alaska, had a very aggressive plan to recover heat from a lot of the tube-operated terminal equipment to help maintain the temperature in their facility and reduce heating costs.

Various forms of geothermal heating and cooling are also available but there are many variables, which include water tables, quality of water, overall installation, and maintenance costs of such a system, and the recognition that by strict definition, heated water becomes a regulated waste stream because heated water pumped back into the environment can have significant impact. For groundwater, the impact may or may not be minimal; for water pumped back into bodies of water, the impact on the entire water body's ecosystem can be significant. We will come back and look at some additional opportunities for capturing "heat" associated with any number of operations. When we start talking about heat recovery and air distribution, something as simple as effective ventilation allows the recovery of a tremendous amount of heat value in production and related operations that, in turn, can help to reduce the various costs of heating.

Cooling offers fewer opportunities today simply because of the laws of physics and the current state of technology. However, geothermal cooling and simple wet cooling properly deployed in appropriate climates offer opportunities that are seldom considered. With the rising cost of fossil fuels and constantly increasing regulations, and the economic cost of regulatory compliance, these are areas that are also going to need further study. One big issue with cooling when one gets beyond a certain facility size is that many current techniques use chemicals that are hazardous, and then they become a challenge to manage, represent increased liabilities, and, in most cases, create hazardous waste streams; ammonia is a leading coolant.

For very large facilities or for larger properties such as schools or school campuses, large office parks, prisons, ports, and similar operations, it is not unusual for the occupying entity to build its own power generation facility and, for larger properties, water treatment and sewage treatment facilities. Based on current technology, new forms of energy generation, including such things as pebble bed atomic reactors, might make sense. Many power generation processes depend on the heating of water and that represents an additional set of risks liabilities and waste streams. One thing often overlooked in the development and management of heavy industries/large commercial facilities are the costs, dangers, of waste streams associated with boiler water treatment chemicals, scrubbers and other peripherals, and, support operations that are required to maintain such a facility.

4.2.1.5 Utilities and Space

Except for fairly large facilities, utilities are most often purchased so there is little in the way of hazardous materials or waste streams, especially waste streams of airborne or waterborne emissions. That leaves the issue of space. This becomes another set of balance or trade-off equations should one build up or source, can production or other inherently dangerous processes be isolated from or integrated into the overall space? How does building design and construction impact "livability" and adequate physical working space? Can effective use of any number of different techniques reduce the carbon footprint, the physical footprint, or the environmental footprint of a facility? One area that will receive more attention is the use of pervious rather than impervious materials for parking spaces. Another will be the use of natural light to supplement electrical lighting; yet another is proper siting on a lot and/or proper selection of a lot to allow for proper siting and the need to plant or take advantage of existing tree lines based upon heating, cooling, and natural weather patterns. In essence, we are saying is that every way you can reduce the need or use of generated power is a way to improve your bottom line while at the same time providing for a more environmentally friendly operation. Since we are taxing on the actual impact on the environment this is not an area that should be ignored although in many cases it may be an area that falls outside of the organization's ability to impact change.

4.2.1.6 Cleaning Supplies and Disposal Methods

Most people today recognize that the range of cleaning products, and, to a lesser degree such things as oily rags, represent another area of risk and environmental contamination. There are far too many products that carry far too many risks for us to cover in this rather high-level overview. The primary issue here is that a very large number of cleaning supplies represent hazards of one sort or another, and there is an even larger danger in mixing the miscible or chemically reactive cleaning materials. Going into detail, many facilities that require rotating machinery and use petroleum-based lubricants do not throw away oily rags and other materials that absorb lubricants or solvents. There are a host of companies today that will either manage or recycle such industrial effluvia. The important point is to recognize the inherent risk associated with the materials and then manage that risk. (See some of our discussions to follow.)

In summary, there are far more hazardous materials and possible hazardous waste streams in the typical office environment and in the janitorial arena than most businesses ever consider, and most business managers do not give enough thought to this. When one starts looking at the incremental volumes and risks from all operations from the simplest office operations to the most complex production operations to the most difficult process maintenance procedures, the total volume of materials that need to be examined is huge. Chemicals and substances of all sorts are subject to regulation and therefore must be considered or treated as "regulated" materials. All are subject to some form of control, and when all areas are recognized, both the number and the quantity of such regulated materials grow quite large. That does, in fact, represent a measurable and quantifiable risk to the organization.

4.2.2 B Logistics Functions

As one who considers himself a logistician, I take an extremely broad view here, and while some may wish to break out of logistics into supply chain management, transportation, logistics engineering, operations management, and other terms, I still base my use of the word on that very broad concept of logistics as originally addressed by Military Standard 1388, originally recognized by both the United States and United Kingdom. While the standard is not seen in the United States today, the basic concept of Integrated Logistics Support was captured in this standard. The widely accepted definition of ILS includes 11 elements:

- Reliability engineering
- Maintainability engineering and maintenance (preventive, predictive, and corrective) planning
- Supply (spare part) support (e.g., ASD S2,000M specification)/acquire resources)
- Support and test equipment/equipment support
- Manpower and personnel
- Training and training support
- Technical data/publications
- Computer resources support
- Facilities
- Packaging, handling, and storage and transportation (S&T)
- Design Interface

Key elements from that standard—logistics support analysis, reliability-centered maintenance, and configuration management, to name just three—go by various names today and have been integrated into all sorts of automated products. They, and most of the other 11 elements, form the basis for a tremendous amount of modern business activity in both heavy and light manufacturing, and they certainly play a major part in the design and delivery of large systems, including power-generating systems, petrochemical cracking plants, and advanced ocean liners, locomotives and airplanes, or even something as simple as a military side arm.

4.2.2.1 Product Selection Criteria Going Beyond Pricing

Once the designers and engineers have begun to develop an actual physical product, even in the preprototype stage, a logistician needs to get involved. The logistician needs to understand concepts from many different fields including materials engineering cost-benefit analysis, reliability, and, of course, corporate risk, because of the inherent hazards unique to a selected input stream. There is considerable room for new types of collaboration between the design team and the procurement specialist. Procurement specialists are going to have to gain new skills, and design engineers are going to have to give up some old skills in order for the system to function most effectively and efficiently.

Product selection must balance out additional unique factors once form-fit functions have been addressed. Up-front acquisition costs, life cycle costs in terms of maintenance repair liability, and hazardous risk—wherever they are encountered during production or all the way down to the end users if these are private individuals at the retail or home-use level. This also includes considerations related to life cycle disposition costs, the concept belonging to the relatively new field. In 2010, of reverse logistics. Up-front costs or purchase price may be identified as one of the largest drivers, but it is no longer the only driver in selecting component parts. In many cases, price may be more important than the areas mentioned previously, but in other cases higher priced items may prove to be more costly in terms of product life operating and maintenance costs as well as product disposition costs. Society today, through international law and trade agreements, is beginning to levy a charge on the manufacturing or producing organization that develops more perceived risks to the planet.

A brief discussion of three commonly applied plating processes demonstrates why this is truly a critical area today. Please look at the opening news item at the very beginning of Chapter 1. Three standard industrial plating techniques are summarized below. Depending on perceived application and the technical needs, it is in fact the logisticians, as well as the risk management or environmental health and safety personnel, who need to understand the direct and indirect societal environmental costs of choosing component pieces, parts, and fasteners that have been subject to the various plating techniques. This is an issue that needs to be addressed between corporate management, finance, engineering, sales and marketing, and, logistics. This is not a one-size-fits-all or a single department decision-making process.

Cadmium plating (or "cad plating") offers a long list of technical advantages such as excellent corrosion resistance even at relatively low thickness and in salt atmospheres, softness and malleability, freedom from sticky and/or bulky corrosion products, galvanic compatibility with aluminum, and, freedom from "stick-slip." Stick slip is important because it allows reliable torquing of plated threads. Cadmium plating can be dyed to many colors, has good lubricity and solderability, and works well either as a final finish or as a paint base. If environmental concerns matter (*and they do today*), in most situations cadmium plating can be directly replaced with gold plating as it shares most of the material properties. The major drawback to gold plating is that it is more expensive and cannot serve as a paint base.

Zinc coatings prevent oxidation of the protected metal by forming a barrier and by acting as a sacrificial anode if this barrier is damaged. Zinc oxide is a fine white dust that does not cause a breakdown of the substrate's surface integrity as it is formed. Zinc oxide, if undisturbed, can act as a barrier to further oxidation, in a way similar to the protection afforded to aluminum and stainless steels by their oxide layers. The majority of hardware parts are zinc plated, rather than cadmium plated.

Chrome plating. Typically when we are using this term we are referring to what is colloquially referred to as decorative plating—think of chrome parts on a show car or bathroom and kitchen fixtures. This form is typically a 10-μm layer over an underlying nickel plate. When plating on iron or steel, an underlying plating of copper allows the nickel to adhere. The pores in the nickel and chromium layers also promote corrosion resistance. Bright chrome imparts a mirror-like finish to items such as metal furniture frames and automotive trim. Thicker deposits, up to 1000 μm, are called *hard chrome* and are used in industrial equipment to reduce friction and wear.

A brief review here because of the growing environmental concerns about cadmium. At one point in time asbestos was used universally as an insulator, which included such everyday items as potholders and flooring tiles, to name just two. When DDT as an insecticide was first introduced, it changed farming in such a way that a much greater portion of the world's population could be fed from the same acreage of crops. For a number of years carbon tetrachloride (CCl_4) was used as a fire suppressant and a solvent to clean electronic contacts, among other things. Mercury was used for thermometers and many electrical applications, and as a primary material in the formation of dental fillings (many of us remember our dentists turning pennies into "dimes" using mercury). Halogens and other chlorofluorocarbon (CFC) materials led to the introduction of automobile air conditioning and proved extremely valuable in fire suppression systems for spaces packed with computers and other electronic equipment. Halogen was a major step forward for fire suppression for electronic spaces because all the other then current suppression techniques made the very equipment they were designed to protect unserviceable and/or led to significant degradation of the electronics. Two growing controversies in 2012 are use of pervious surfaces, primarily for parking areas and lawns. Yes, that is right—lawns. There is a growing body of evidence that in America, at least, people's love for green lawns is significantly reducing water quality in freshwater bodies and has destroyed large portions of once-rich shellfish and other marine life in rivers, lakes, and streams. The Hudson River, which, for good portion of its length, is the geographic dividing line between New York and New Jersey, was once known for its large oyster beds. In New Hampshire, the Great Bay and the Oyster River were also once homes to large shellfish beds, hence the name Oyster River. Shellfish were essentially wiped out in both these bodies of water a long time ago, and significant efforts are being made today to clean up the water to the point that commercially viable shellfish beds can be brought back. You will see the same issue along the East Coast of the United States, extending beyond shellfish, although some of the causes are quite different from the nitrogen-overload created by fertilization to give us green lawns. Today, we recognize every one of these as presenting significant hazards to our individual health and the health of the environment; many of these materials—or chemicals, to be more accurate—are either no longer available commercially or are subject to strict and very limited use applications.

Another issue is that technology has demanded the use of more precious metals and other rare elements. Many of those are mined and recovered from a shrinking number of geographic locations in fewer and fewer countries; there is growing concern that this is both an economic and defense weakness in the global economy as it is allowing nation states to attain monopolies on critically needed materials. For those interested in reading more about this subject, CRS report R41744, "Rare Earth Elements in National Defense: Background, Oversight Issues, and Options for Congress," dated April 11, 2012, gives an excellent analysis.

Hopefully, all this drives home the point that understanding the industrial processes behind components as small as an individual screw and as large and complex as complete electronic/computer printed-circuit assemblies, which go into a product, is critical to making many decisions. Those decisions include what parts to specify and what parts to make in-house versus what parts to outsource and how design development and delivery of products impacts the environment and reflects corporate culture, mission statements, and goals. While, at the same time, decisions about where to build, how to build, and the need to build also have a direct impact not only on the environment but on the risk and liability of the "owning," or in many cases the leasing of a commercial facility, whether that be a manufacturer or a law firm.

4.2.2.2 Transportation Packaging

We will use slightly nontraditional terminology here in the hopes of maintaining clarity for all those who are not packaging professionals, and are not interested in becoming packaging experts. Technically, everything that contains, restrains, protects displays, promotes, and adds to product usability or value at the user purchase level is "packaging." In a more focused framework, packaging is most often described as serving four functions and falling into three groups. The three groups that are generally used to categorize the entire packaging universe are: primary packaging, secondary packaging, and tertiary packaging. We are going to reduce that to two broad areas: first, *primary packaging*, that which occurs as part of the production process, and in many cases as an integral part of production. Examples of the latter are individual packets of salt, ketchup, or mustard and the individual dosage packaging of both over-the-counter and prescription drugs. Here, packaging is essentially an integral part of the product. Second, we consider *transportation packaging*, that which is applied, or conversely removed, at the shipping and receiving areas of organizations.

Primary packaging may be an integral part of the production process or may occur at the very end of the production process so that it is an integral function of the product characteristic. Quite frequently, initial packaging is literally contiguous to the end of the production process, so it will be addressed in the final segment (as soda, beer, soup).

Logistics is involved with transportation and distribution so transportation packaging will be discussed here. As part of the overall product development cycle and design process, there are issues that need to be addressed with logisticians and packaging specialists throughout the entire product development process. These decisions will directly impact on choices about specific technical functions for production line configuration, product package, product packaging, and transportation packaging.

Transportation packaging has one overarching mission: preservation of the product from point of manufacture to the customer. This packaging adds to the value of the product in many direct and indirect ways. This includes the shipping container, interior protective packaging, and any unitizing materials for shipping. It does not include packaging for consumer products such as the primary packaging of food, beverages, pharmaceuticals, and cosmetics. In the United States, transportation packaging represents about one-third of the total purchases of packaging; the balance is attributable to primary packaging for consumer products. To design a transport package one must have goals or objectives in mind. These will vary with products, customers, distribution systems, manufacturing facilities, etc., but most transportation packaging should address the following:

Ease of Handling and Storage—All parts of the distribution system should be able to economically move and store the packaged product.

Shipping Effectiveness—Packaging and unitizing should enable the full utilization of carrier vehicles and must meet carrier rules and regulations.

Manufacturing Efficiency—The packing and unitizing of goods should utilize labor and facilities effectively.

Ease of Identification—Package contents and routing should be easy to see, along with any special handling requirements.

Customer Needs—The package must provide ease of opening, dispensing, and disposal, as well as meet any special handling or storage requirements the customer may have.

Environmental Responsibility—In addition to meeting regulatory requirements, the design of packaging and unitizing should minimize solid waste by any of the following: reduction, return, reuse, recycle.

Since transport packaging should always be economical, the above goals should be balanced or optimized to achieve the lowest overall cost (McKinlay, 1999).

In 1914, American railroads, which at the time were carrying most of the freight in the United States, recognized and authorized the use of corrugated and solid fiberboard shipping containers for packing many different types of products. Motor carriers, in turn, followed the railroads' example in 1935 when they adopted their own packaging rules that often called for fiberboard boxes. This standard packaging, employed during the early months of World War II, failed colossally with a critical impact on the amphibious landings at Guadalcanal in 1942. This demonstrated how critical proper packaging design really is, forcing the military to conduct their own research and create their own organizations to address packaging. This eventually led to military specifications (MILSPEC) and military standard (MILSTD) packagings. The universally accepted packaging available at that time had proven adequate for rail and highway movement within the 48 continental United States but was grossly inadequate for the movement of goods overseas, into different and changing environments, and under prolonged exposure to the elements (Maloney, 2003). The scope of this disaster and the lessons learned led to the establishment of the School of Military Packaging Technology; the school disappeared from Aberdeen Proving Ground in October of 2009 as a result of BRAC, and some of this mission was transferred to the Defense Ammunition Center. Part of the reason this occurred was the rapid change in technology, specifically materials technology and packaging technology, which raised the overall standards for commercial packaging as well as military packaging.

The three biggest threats to product and packaging from a transportation and distribution standpoint, and in descending order, are moisture, temperature, and pressure. If retail unit packages, which may contain multiple quantities of individual products or multiple packages of multiple quantities of products, include sufficient protection against moisture, there is less need for either more expensive or additional transportation packaging. On the other hand, as an example, if the organizational concept requires all outbound shipments to be properly shrink-wrapped with the appropriate shrink-wrap material, then theoretically the need for vapor barriers and/or desiccant materials may be reduced or eliminated. That might be particularly true for products that are produced for a local or regional consumption yet still in relatively large quantities. This is why logisticians need a broader background than most recognize, and why logisticians need to be involved very early in the product development stages. Most texts on packaging identify four primary functions, although they may use slightly different terms or variance. The four functions of packaging are to contain, protect, facilitate handling, and promote sales (Ramslund, 2000).

One area worth discussing briefly is packaging materials and packaging quantities. For those who entered the workforce in the 1960s and 1970s, and have been involved with business-to-business commercial and industrial products, it comes as no surprise that packaging is a major issue. For those who study retail markets, consumer behavior, demographics, and psychographics, it is not news that packaging at the individual unit level for retail products presents serious challenges to getting the right balance between product protection and preservation versus safety versus ease of opening. The need to balance all three issues presents a myriad of challenges and has led to well documented consumer frustration and dissatisfaction, while raising product costs as well as waste stream growth.

The many issues related to movement of goods from point of production to point of sale is another area that represents general waste issues, damage and loss issues, and environmental issues and impact. While not really addressed here, the whole new field of reverse logistics applies the moment goods leave the point of production. The more important point here is that the concept of reverse logistics must be engineered into the product, the production line, and the packaging function. This is necessary to address the growing legal ramifications associated with the production processes that generate hazardous waste streams and emissions, as well as the ultimate ownership and responsibility of goods that are, or contain, hazardous materials.

The types of material used for what is known as secondary and tertiary packaging, the methods used to secure or "build" the packaging, and, the methods, processes, and materials used to generate visual information on the packaging all can add hidden costs that represent hazardous materials exposures and increased packaging disposal costs. Some examples of specifics would be: adhesives used in building the physical package, or methods of closure/sealing; inks, paints, or other materials applied to the packaging to provide visual appeal or critical logistics and transportation information; preservation of the package itself; application of fumigants to tertiary packaging and containers; and passive and interactive RFID tags.

One often overlooked area that this text will not address in detail but is worthy of separate discussion is invasive species and/or environmental issues related to the movement of goods by sea from one ecosystem/environment to another ecosystem/environment that is geographically removed by great distances.

Kudzu, Asian carp, Zebra mussels, and long-horned beetles are just a few examples of invasive species that have drastically changed or are currently changing local ecosystems. The most egregious examples of such unplanned consequences from modern air and sea transport can be found in Guam, unfortunately one ecological disaster area that very few Americans, including business people, educators, legislators, and logisticians, are aware of.

It is now generally accepted that Americans unknowingly and inadvertently introduced brown snakes onto Guam in their post–World War II efforts to rebuild the island's economy. Those brown snakes have wiped out all native birds on the island, and it is now a full-time and not inexpensive job to manage and/or reduce the snake population on the island; there are ongoing efforts to restore an avian population to Guam but it is an expensive and extensive long-term proposition.

While less pressing, those issues apply to commercial and noncommercial movement of goods and the transport devices (including those used strictly for pleasure) between areas within the United States. At the current time, because of an invasive species in parts of Worcester County, Massachusetts, raw forestry products, including firewood for personal use, cannot be moved out of the county for any purpose (such as camp firewood) to another area. If you take the time to look at signs posted at many public boat ramps throughout the United States, you will notice that there is both a request, and a legal requirement, to properly clean or purge boats when they come out of the water even for movement within the local areas within the United States. An overlooked partner in this area is USDA APHIS. Agricultural products may not freely move into and out of every state because of the organisms they might carry whether that is some form of plant disease such as blight or some sort of living insect such as arachnids or nematodes. Arachnids and nematodes love to travel in/on untreated wood, so we now have international phytosanitary standards. In the past, treatment of wood products was (or in some cases still) uses what we now recognize as extremely toxic/poisonous materials as an alternative to other forms of treatment; recent cases demonstrate that under certain conditions the treatment of such packaging materials used in transportation, such as pallets, dunnage, and blocking and bracing can contaminate the product being transported.

As you can see there is a lot more to product packaging, packaging for transportation, and transportation itself that needs to be considered and discussed. We will come back to that later, but now let us move on to packaging directly related to the product rather than its shipment.

4.2.2.2 Material Handling Equipment and Material Handling Systems

If you have material handling equipment, then you have hazardous materials issues as well as environmental health and safety issues. This is not a bad thing, but, if you have not properly addressed all the requirements or you simply did not understand that there were so many requirements, this could be a weak area for many operations.

All such items use some form of motor that is rotating machinery, so immediately there are personnel safety issues and lubricant issues. Internal combustion engines, regardless of the fossil fuel, create additional series of challenges based on the fuels themselves and air monitoring requirements and circulation considerations. Electric powered units have batteries. Depending upon what kind of batteries are being used, battery charging and servicing stations as well as other unique EHS requirements that might include acid absorbent spill kits and neutralization materials, eyewash stations, open flame and heat restrictions, and possibly special fire suppression and fire-fighting materials and protocols, are needed. All represent additional costs, risks, and exposures. Both industrial and small lithium-ion batteries have been known to cause fires, so adequate research needs to be done both on the batteries themselves and charging systems that will be used, including the possibility for the need of fire suppression and current limiting devices. If you are using lead acid batteries to start your material handling equipment, then you need the eyewash stations and servicing stations with adequate ventilation and personal protective equipment. For an operation using rotating machinery, including typical MHE, lubricating products, as well as lead acid batteries, are part of the package. There are two different sets of spill response/control items and two separate sets of requirements for safety materials and personal protective appointment.

The larger issue is recognizing what hazardous materials are inherently parts of the material handling equipment. One category of hazardous materials must be kept on hand to service and support the material handling equipment, and whatever sort of waste streams you are generating. There is a second set or group of materials that represent recognized and hidden waste streams. This includes everything from coolant to battery acids drained or unused excess oils, and, of course, oily or otherwise contaminated rags. Solvents used to clean the equipment itself, as well as the tools in the work areas and various preservatives, may need to be treated as hazardous materials. Many service and maintenance aids, once they have been opened/used, that might not be treated as hazardous materials initially will have to be treated as hazardous wastes when disposing of them. That includes even the smallest quantities in the bottoms of cans and drums, or tubes, swabs, or brushes that have any foreign materials adhering to, or absorbed in, them. Some of the materials used in operating and maintaining material handling equipment are essentially incompatible, which means they must be segregated and that special efforts must be made to properly segregate those waste streams as well.

4.2.2.3 Transportation Methods

Most purchasers of transportation services, and the vast the bulk of the transportation operators, since the majority are single mode operators, usually look at the cost–time convenience factors when making decisions in regard to mode selection. Today, informed businesses who recognize the impacts of life cycle management are also looking at other costs and impacts in deciding on their total distribution chain, even if that function is exclusively provided by 3PLs and outside contractors. Energy efficiency and CO_2 emissions are both factors that need be applied when designing the distribution system. A properly designed system can minimize costs to the company in relationship to inventory cost; convenience for the customer, or rapid delivery; and green concepts or impacts. The chart below clearly demonstrates that there are a number of factors that need to be considered by both transportation sector providers and users when making decisions in regards to transportation and distribution models. See Table 4.5.

For many reading this, a lot of the information in the table is superfluous to the point of overkill, but for certain industries and different specific businesses and sectors, all of the information has value and impact.

Realistically, very few modes of transportation can provide for uninterrupted door-to-door movement of goods from place of production to place of sale or use (the two are not necessarily the same when dealing with retail products). That means almost all goods will make some part of their journey by truck. That by no means suggests that truck transportation is the best, or even a good, choice. In addition if one understands the transportation industry even for a material that travels exclusively via truck, it is likely that the material will flow onto at least three and possibly as many as seven or more different transport units before arrival at its final selling point. Many

Table 4.5 Modal Comparison of Environmental, Safety, and Health Factors

Mode	*Energy efficiency index*	*Speed*	*Average haul*	*Deaths per billion passengers (kms)*	*Date introduced*	*Vehicle life (years)*	*CO_2 emission (gms per ton per km)*
Air	1	400	1000	0.02	1958	22	540
Truck	15	55	265	2.4	1920	10	50
Rail	50	20 (200)	500	0.55	1830 (1970)	20	30
Barge	64	5.5	330	Very small	17th C	50	
Pipeline	75	4.5	300	Negligible	1856 (1970)	?	
Liner	100	16.5	1500	1.0	1870 (1970)	15	

Source: Alderton, P., *Port Operations*, Informa, London, 2008.

producing organizations capture the major issues related to operating their own fleets or contracting out to 3PL or other external providers. What few recognize are the huge chemical costs related to the service and maintenance of a vehicle fleet. All information that applies to material handling equipment shown in Table 4.5 applies to an organization that wishes to operate its own vehicle and/or decides to maintain such a fleet.

It takes a careful examination of the information in the table to make decisions. And for many, the decisions are going to be driven by other factors. It is only when we start aggregating large quantities and talk about initial distribution that a lot of this information comes into play. Nonetheless, it is critical that the issues represented by the information in the table be considered when making large-scale, and to a lesser degree, smaller-scale or local transportation mode decisions.

Many travel booking services now provide CO_2 emission figures as part of the reservation package for air travel. This is shown as how many tons of CO_2 emissions each flight leg of a trip represents. Given the direction of global regulation and taxation, the concept represented by that information now carries an increasing direct economic burden, whether that is viewed as a hidden cost and whether or not this is passed on to the ultimate consumer of the good or service. There are inherent safety and cost considerations related to each mode, and based on either, it is fairly easy to create ordered lists to help choose transportation modes for the movement of goods. Reliability is another parameter; however, two parameters are becoming more important: energy efficiency and emission generation. With careful planning and attention to detail, the supply chain manager can build on improved energy efficiency and reduced emissions to reduce the cost to the producer and/or the consumer. At the same time wise choices and careful planning within the supply chain also contribute to the reduction of environmental impact. In the end, based on current trends, society as a whole is going to have to want to understand and then pay for the economic impact of the various modes of transportation, as well as the economic impact of the various modes of production and packaging. Businesses are going to have to accept and understand the negative impacts of industrial processes, which include production and transportation, upon the environment and then decide how to either avoid or allocate those costs.

While there are no absolutes, some basic decision points can be presented here. For those concerned strictly with local operations, the chart above has little practical application. For those who are shipping strictly within a region, the chart begins to take on meaning, but the options and choices are still limited. For those distributing nationally or globally, the chart begins to take on a lot more importance. Realistically, both those who are shipping and shipping companies themselves need to pay particular attention to the first three columns and the last column in the chart. In addition to that, those companies who are distributing beyond the region must factor in the issues of warehousing and distribution centers. Today, the majority of business texts treat the transportation distribution model as a trade-off between transportation costs and distribution center/warehouse costs and convenience. The big issue is the cost of inventory sitting in warehouses and distribution centers, as well as cost of inventory in transit. That is an excellent model and is still the foundation for such discussions. Today, however, driven by the growing body of scientific research and the increasing tendency of governments to regulate and impose "carbon taxes" domestically as well as globally, it behooves both the business community at large and the transportation community at large to look more carefully at some of the information provided in Table 4.5.

Today, the final decision must be driven by a combination of company culture; interest in, and legislation encouraging, "green" approaches to business; consumer perceptions; and cost benefit analysis that trades off time versus delivery cost versus convenience. Make no mistake, for any entity that is delivering large quantities outside of its region, all those issues must be taken into consideration.

4.2.2.4 Transportation Providers and Protocols

If one looks at inbound logistics for goods-producing organizations, time cost, liability, and reliability all become drivers. Often, adjusting overall production schedules may allow for a move to less expensive and less environmentally unfriendly modes of transportation. That is a key concept in this rather short section. The just-in-time approach to manufacturing or production does not necessarily or automatically negate the use of less expensive modes of transportation. Typically, less expensive modes of transportation also represent those modes with the lowest carbon footprint and environmental impact. The trade-offs include time as an absolute, reliability, and seasonal impact on the mode of transportation. Something as simple as scheduling pickups or deliveries at the manufacturing facility for "off" hours can have significant direct and indirect environmental and economic impacts. Time of day pricing for surface transportation is now embedded in the Southern California port regions and can be expected to expand throughout the United States in all major port areas.

Commercial trucking operations and rail operations can have significant impact on congestion as well as pollution. If we parse the normal 24-hour day into three shifts, we could make a case for shipping and receiving operations being restricted or focused upon what is traditionally labeled as the "midwatch" by the military. That could be an arbitrary 8-hour band of time. That might be from 8 pm to 4 am. Taking the theoretical step further, shipping and receiving areas can be physically isolated from the rest of the facility. Then there might be additional heating and cooling cost reductions associated with operating such facilities when the rest of the facility is closed down. This is not to suggest that any work has been done to quantify the advantages and disadvantages of such a concept. Humans are "wired," if you will, to operate during the day and sleep during the night, so safety in terms of driver ability/capability is one of the many factors that would have to be looked at under such a scenario. From a pure transportation cost basis and a pure environmental impact basis, there are a tremendous number of positives to exploring such an approach.

In addition to mode selection and time of day decisions, another major decision for many organizations—one that should help drive the outsourced transportation function decision—is selecting an appropriately equipped and operating carrier. Here, the issue is less a hazardous material or waste material consideration than it is an environmental impact safety and reliability issue.

The last component of the transportation issue is packaging. While not a major consideration, the total distribution chain for product will drive packaging design and packaging decisions. Proper choice of primary and secondary packaging, coupled with the intrinsic characteristics of the product, may allow the product to move equals the reliability and safety in any mode of transportation. On the other hand, larger industrial products which might not ordinarily be considered "fragile" might need additional packaging for ocean voyages especially since today a very large number of such items are "modularized" so they may fit into shipping containers.

4.2.3 Production Functions

4.2.3.1 Lubricating and Cooling Chemicals

This is the area that creates the greatest opportunities and the greatest challenges. Some of the issues related to lubricants include purity, compatibility, particulates, temperature, and effectiveness of the lubricant. To begin with, let us look at a hypothetical situation. A different lubricant that is 10% more expensive can increase cutting tool life by 20%. In doing so, it also reduces total volume of lubricant by 5% but because of its chemical makeup requires additional monitoring and special handling for disposal, one would have to look at every one of these issues, do the cost-benefit analysis, and consider some areas that were not mentioned before making a determination on switching to the new lubricant. On the other hand, if we have a lubricant that is 30% less effective but results in the production of a nonhazardous waste stream, which therefore would not be a regulated waste stream, what would be the impacts on total costs? One of the problems here is that many of these costs can never be attributed directly to a finished product and fall into the very large category of indirect costs or overhead costs. Notice there are no easy or even obvious answers here. One of the drivers—if incremental costs are small enough and once all the calculations are done—will be the corporate culture and the corporate climate as well, as the current regulatory climate.

Sticking with lubricants, there are additional considerations for those facilities that are consuming large quantities of lubricant. Typically, if we are talking about large enough quantities to use in a closed-loop application, we are talking about processes that generate particulate as well as miscible and immiscible wastes, while generating some amount of heat. In terms of recovering "value" from such lubricants, there are a number of additional issues that need to be taken into account. The actual weather in the geographical location of the facility may become a major determinant in all of those considerations. If large enough amounts of lubricant are being used and the processes raise the temperature of the lubricant sufficiently, it is quite possible that those lubricants can be used to help maintain ambient temperature within all or some part of the facility. A technical question in such a scenario might be how to effectively filter out the scrap materials—typically metals—as they represent a number of distinct and separate challenges in a circulating system. A second consideration in addressing heat energy recovery is what to do once the heat energy from those circulating streams has been extracted, captured, or otherwise used to reduce energy costs to the organization. Assuming for the moment that a circulating system can handle the lubricants with any suspended solids including metals, capturing that stream as a source of "BTUs" and injecting or mixing it with the primary fuel source at the facility can yield additional benefits.

If the furnace or incinerator is operating at a high enough temperature, some of the hazardous characteristics and whatever contaminants a lubricant is carrying may be mitigated or eliminated. Carrying this one step further, the use of a high-temperature incinerator mixing various waste streams of varying BTU values can essentially eliminate or remove hazardous characteristics, while the energy released may be used to drive or impact another process; that could be anything such as enhancing the catalytic reactions or providing hot water in the bathrooms. On the one hand, some of this is fairly esoteric; on the other, there are many practical applications that businesses have used and are using today. The end result is to convert as much hazardous waste, and to a lesser extent nonhazardous waste, into useful applications, reduce overall negative emissions and discharges, and reduce the absolute volume of solid waste, while, at the same time, recovering every BTU of energy generated in the production/manufacturing process.

A separate and related issue is the move towards less hazardous and/or environmentally friendly or neutral lubricants. A good example would be replacing petroleum-based oil with a vegetable oil. The petroleum-based oil in its waste form must be treated as a hazardous waste, and it represents a real if low-level risk to the environment. Vegetable oil, on the other hand, is a natural biodegradable product that represents no hazard other than sheer volume and would not be subject to regulation unless it met some other criteria such as higher temperature.

The last issue related to lubricants, and all other liquid vapor streams, is the fact that some cases covering heat energy, which means effectively cooling lubricant, reduces its inherent hazard and/or may shift the hazard classification. As we pointed out earlier, material that exists at an elevated temperature is, by definition, a regulated hazardous material if it has to move off the facility.

4.2.3.2 Fastener Methods and Products

Essentially there are two ways put materials together: mechanical fastening or adhesives. A good foundation in materials science would help many logisticians and probably many other senior business managers, as well as the engineers and scientists designing products.

Metal fasteners such as nails, screws, staples, clips, and rivets present their own challenges in terms of space, weight, cost, and aesthetics. In many applications, use of metallic fasteners either requires visually appealing materials or materials that will not rust or corrode. Another issue for many threaded fasteners is whether they will go into materials where excessive torque will be applied for maintenance and repair, or for that matter on the part of retail consumers, such over torquing might strip out the threads in soft materials. Heat conduction and electrical conduction are other issues that need to be factored into the equation when determining fastening decisions, when building products, and for that matter when designing certain types of packagings. From a practical standpoint in mass production, fairly precise automated means are being used with competition many of these assets the possibility of malfunction leads to scrap, and in many cases we are talking about increased use of energy to accomplish the task. Typically, that increased energy is probably offset by the savings in labor costs, but for large enough operations, those issues need to be examined at least once.

Adhesives present a whole different set of challenges. The vast majority of commercial adhesive applications are dependent upon chemical characteristics and chemical reactions; in some cases they will also require pressure and/or heat. Generating pressure or heat means using more energy, which increases the use of power, so again the carbon footprint for the operation/product is going up. Considering both the aesthetics issues and the production need to bond dissimilar materials, adhesives inherently offer many advantages and many choices. The challenge comes in having expertise on staff or on contract so you are able to select the most effective combination of adhesive application and setting methodology, as well as the inherent risks and worker exposure considerations with the use/application of the adhesives themselves. In simplest terms, there is a great deal of expertise necessary to evaluate adhesives in order to make sure that they are either the least hazardous, or, if at all possible, not hazardous. That process offers the greatest life cycle advantage in costs, which are rarely recognized but still significant.

4.2.3.3 Surface Finishing Products and Processes and Other Sealer Treatments

A number of issues here include aesthetic considerations, finishing processes, application techniques, energy consumed, UV light transmission, IR light transmission, ability of users to grip or grasp the product, refractivity and visibility on the finished surface, and heat and pressure characteristics. This last is particularly important when talking about the myriad consumer products that invalue the interface of touch-sensitive surfaces. Let us look at some of the more important issues when it comes to hazardous materials and life cycle costs. The application of paint lacquer and sealants usually entails the use of a material that has a complex chemical character and typically represents a regulated material. That means you have all the costs, risks, and liabilities associated with procuring, storing, and using the material, plus very specific OSHA requirements, but perhaps more critical is the decision on how to apply such materials. To some degree, there is a trade-off between convenience, cost, personnel safety, and, waste streams and waste minimization and/or processing costs. If one uses a "dipping" process to apply a finished coat to a product, that represents one set of risks and environmental considerations. If one chooses to use a "spray" technique, that introduces an additional set of considerations and now means you must take into consideration moving from a solid or liquid state into an airborne state. So there is an immediate increase in the sheer volume of material used and the space, in cubic measurements, required, as well as a much more complex and physically larger recovery system. That does not mean one technique as an automatic or inherent advantage over the other. It means that you need to understand all the countervailing issues to make wise decisions before you are able to start production. In contrast, if the best, safest, and easiest-to-apply finish is too reflective, people will not be able to see what is on the surface or what is immediately below the surface.

Success in this area is dependent upon access to a combination of expertise in materials and materials science, and expertise in current and looming statutory requirements and legislation.

4.2.3.4 Chemical Processes, Chemical Selection, and Chemical Recovery and Treatment

There are two primary topics for discussion here. First is the choice of completing processes in-house or outsourcing; the second is the selection of products and processes to be used for component parts and/or possibly assembly and internal application of some of the most common processes. Most business majors, managers, and owners who feel no need to understand industrial processes such as printed circuit manufacturing and various plating processes and options or spray painting versus dipping or other process, are making a mistake, given today's regulatory environment and social shift. Any takeover, expansion, or new business must reflect awareness on the part of the management team of the inherent risks and advantages of such kinds of industrial processes. The same holds true for logistics managers in existing operations; they must understand the inherent risks, vulnerabilities, and life cycle impacts of such industrial processes.

Printed circuit assemblies, whether they are called boards, strips, or films, essentially depend on an acid etching process that removes copper from a coated surface. Medically, copper is recognized as one of the many heavy metals. It is critical to the human body but it is also toxic to the human body and therefore the more lessened copper waste is in any form, including dissolved in acidic liquids, the better the environment. Unfortunately, the reality is that copper is the most effective and efficacious conductor of electricity, so we will continue to use large quantities, for the foreseeable future. That is not necessarily a bad thing but overuse and unnecessary use of copper in terms of industrial processes and manufactured products should be a consideration. While there are no worldwide shortages of copper at this point time is not logical, rational, or safe to assume that will hold true in other 20, 30, or 50 years. If copper becomes less available, the cost of copper will go up.

The acid etching process is the second part of the equation. By its very nature, acid etching consumes large quantities of acid and generates large quantities of waste liquids. The costs associated with the use of hazardous materials are far greater than most recognize, and represent an overhead or hidden cost not easily allocated to individual units of production. The acidic waste must be neutralized or treated; whether that be on-site or off-site. Realistically, the area of waste materials have become too expensive except for the largest operations acid etching procedures are going to lead to fairly significant hazardous waste and disposal costs and use of third parties. Environmental health and safety costs that involve compliance training and facility design are considerably greater than most people realize. An excellent example would be a facility that was going to engage in a new, first-time acid etching process. Today, that would require a considerable amount of personal protective equipment; very specific ventilation and circulation requirements including, by definition, acid-resistant materials as distinct from the traditional galvanized sheet stock that is used for the bulk of circulation systems in commercial and industrial facilities; storage separation and segregation issues, as well as effective fire protection and fire-fighting systems in areas where the assets are used or stored; additional training and training compliance costs; and, an often overlooked aspect, facility design that ensures leaks whether in storage areas—pre- and post-production, or in the production operations—are captured underneath the work spaces and not allowed to run into the ground or run into the normal sewage recovery or storm water systems. As you can see, there are considerable costs, risks, and liabilities associated with the production of printed circuit assemblies in the commercial environment.

Once the assemblies are created, components need to be mounted onto them and soldered. The choice of soldering material and the soldering technique have environmental, economic, and hazardous material life cycle dimensions. Spot soldering versus wave soldering is one consideration, while leaded versus unleaded solder is a second consideration. Do not assume one has an inherent advantage over the other.

The use of plated products is another little understood area. Very few operations today do their own plating; there are two simple reasons: the plating processes today still depend on a chemical process that uses arsenic compounds extensively. This is simply too toxic a material for anybody but experts to be using on a routine basis. That leads to the second reason the cost of adequate compliance the moment one gets involved in the use of such toxic materials comes extremely high. The larger decision beyond where plating should occur is whether there is adequate justification for plating as a technique today. This also tends to be an energy-intensive process that generates very large quantities of waste. That waste often contains precious metals that need to be recovered. The actual plating location now has the daunting task of developing adequate and cost-effective separation tanks to recover materials and then handle what would still be partially contaminated highly hazardous waste materials. For lesser reasons, the use of plated products does need to be minimized across the commercial spectrum. A look at the trends in the commodities market, especially at the precious metals and rare earth segments so critical to modern solid state and high-technology production, will shows that from a national security, and for commercial production processes and many high-value, high-technology products, the number of precious metals, and to a lesser degree other critical raw materials, are no longer available from the large numbers of sources they came from less than 50 years ago. The sources for many of these critical building blocks for high technology products is now concentrated in a very small number of nation states creating a dangerous and possibly very expensive situation. Electrolytic plating is also available as well, and it reduces the use of toxic chemicals such as arsenic. It is a much slower process and cannot produce the same thicknesses. Such processes are more typically used in the decorative arts rather than in high-technology products.

As pointed out earlier, there are many different processes that depend on chemical action and reaction that include abrasion, polishing, and finishing processes such as painting or removing surface coatings from products or simply preparing surfaces for further applications or treatments. There are also issues related to heat and electric conductivity and friction reduction. Proper treatment of surfaces made of the correct materials can reduce or eliminate the need for lubrication.

4.2.3.5 Material Usage and Selection

The previous two topics highlighted many of the issues related to material usage and selection; the two most important take-aways are (a) that one must understand the processes that create component parts of assemblies and/or finished products, and (b) EHS considerations, whether from the external production of purchased parts and components or from internal production processes, have significant financial implications as well as socially responsible marketing and risk dimensions.

4.2.4 Product Functions

4.2.4.1 Need for, and Choice of, Operation and Maintenance Consumables and Replaceables

Proper selection of materials that impact after-sales activities and decisions regarding the need for and the nature of consumables are key issues related to product. One example would be a trade-off analysis examining the use of a less environmentally friendly and more expensive lubricant that lasts for the service life of the product versus the use of a much less expensive and more environmentally friendly lubricant that requires regular user application. The first choice makes the product much more user-friendly and essentially eliminates the possibility of over- or under-lubrication or unintended waste streams. That, in turn, makes the product more consumer-friendly and eliminates possible misuse of lubricants by the user. The second choice represents an improved return for the manufacturer on the product sale and allows for a claim that the product is more environmentally friendly. That would allow the manufacturer to claim a "green" advantage for the product. The downside of the second choice, under a growing body of regulations, is that the manufacturer will be forced to assume some of the liability and responsibility for misuse of the lubricant and it also increases exposure and liability on the part of the purchasing organization. That second choice also increases the quantity of regulated maintenance materials throughout the supply chain. Another example, representing actual product rather than support, would be the selection of the material for a knife blade to include the nature of the material, the hardening processes used for the material, and any plating finishing processes. Again, a choice to go to a more expensive material if it were going to eliminate or greatly reduce the need for sharpening, might be driven by consumer acceptance and societal practices, but there might also be issues related to unsupervised (in other words done by the consumer)—processes that generated hazardous wastes. Yet another example are the decisions as to whether to use batteries, what kind of batteries to use, and whether they need to be charged/serviced by consumers.

Other considerations often missed in product design and development have to do with maintenance and shelf life. Today's society is often characterized as a disposable society. The vast majority of consumer products, especially consumer electronics, are replaced due to consumer preference and technological advances, not because of product failure or end of usable life. The initial product life cycle for many such products is becoming extremely short. As a result there is now a strong and growing secondary or reuse market for those products. This is not a repurposing of the product but merely a redistribution into a different economic marketplace. This raises a number of issues that could usually be ignored for retail products. One half of the life cycle issue today is design and development considerations for resale of products into secondary markets. That involves a better understanding of required intervention on the part of the manufactures or its agents for reuse in those secondary markets, as well as recognition of the extended exposure and liability. The other half is a new or expanded focus on accessibility for maintenance, shelf life considerations, and service life for component parts.

For products as mundane and inexpensive as O-rings and grommets, motor brushes, and lubricants and batteries (where location and accessibility become issues) to other shelf-life items and a host of consumables such as ear cushioning and earpieces, manufacturers now need to factor in the extended life cycle of the product in the secondary market, impacts on the product, and the increased liability/exposure on the manufacturer.

Under this new paradigm, for both the business/industrial marketplace and the retail marketplace, ease of replacement and/or design for replacement become important issues. Under this new paradigm, for both the business/industrial marketplace and the retail marketplace, ease of replacement and/or design for replacement become important issues. At the same time, based on current regulatory trends, the full life cycle costs for all products, including post purchase or final disposal costs for end products and components within end products are becoming another driver for product development because all these costs are slowly being passed back to the manufacturer. The end result is that accessibility to, and choice of, consumable and replaceable components as well as hazardous components must be revisited especially in the consumer/retail marketplace.

4.2.4.2 Product Packaging

One needs to keep in mind the relationship between product packaging and transportation packaging. Sometimes this may appear to be the chicken-or-egg-first paradox. An extreme example is if one is manufacturing finished machined parts and moisture protection is an essential part of the process. Manufacturers and the larger personal firearms gun enthusiast market routinely use a wide variety of preservation products broadly known as Cosmoline. There are number of problems with this concept. Number one: most of those preserving materials are by today's definition hazardous materials or, at the very least, hazardous wastes when they are removed due to fact that they are predominantly petroleum-based substances. As such, they are subject to regulation governing their disposal—even a consumer electronic device owned by a private citizen. Two: they inherently require additional production steps, production time, and production materials. If, in fact, they are hazardous to begin with, they also increase the hazardous materials inventory and associated EHS risks and costs to the producer. Three: such materials also require additional time, effort, and solvent if they are going to be removed properly. That then brings up the question of whether this is a supply chain intermediary responsibility (with all the associated costs and risks), or it becomes the responsibility of the final user. This last option, under today's shifting regulatory

Table 4.6 Packaging Involvement With Product Development

Trigger phrase	*Always*	*Sometimes*	*Rarely*
We initiate the new packaging concept ourselves	44%	53%	3%
New product or package conception by other group	24%	65%	12%
When the company commits to the project	50%	32%	18%
First team or project major meeting	50%	47%	3%
When the product is finalized	21%	26%	53%
Close to product launch	15%	18%	67%

Source: Falkman, M.A., Packaging function, structure defined, *Packaging Digest*, 2001 (August 1).

environment, does place the producer in the position of being held legally responsible for the acts of those whom they have no control over—private individuals. This applies to the unpacking of and use of the product and also applies to the initial user, and it passes down to second- or third-tier users and to the eventual disposal of the product with all its associated packaging materials in the post-initial purchase environment.

On the other hand, today's new techniques and products allow transportation packaging to eliminate or drastically reduce moisture within the package. So, the need for individual item preservation is no longer automatically a necessity as part of the production or purchase chain. Another example for consideration: A producer of small hand-held consumer electronics items develops an amazing and very flexible, but thin, peel-off film that offers significantly higher levels of protection with no residue upon removal. This film also reduces the need for a whole level of packaging, thus reducing total cost to ship these products. The drawback is that this film is extremely flammable and must be disposed of properly as a regulated flammable solid. What is the right course of action here? The discussion above is theoretical in nature; there are many factors involved here, and there is no one single simple, correct answer.

Some key points relative to packaging and the unique issues of product packaging: The first thing to keep in mind is that packaging engineers need be involved in the product development cycle from very early on. While very few business and logistics managers think of packaging as an integral part of the product, research clearly demonstrates that this perception is incorrect. Table 4.6 shows that packaging needs to be introduced early in the process of product development, not late in the process or as an afterthought or as a separate activity.

4.2.4.3 When Does Packaging Get Involved in New Projects?

To help put this in perspective and demonstrate how critical packaging is, here are some numbers from a very well-known beer company: In one year, the company spent roughly $2.8 billion on packaging materials and only one-fourth that much on brewing materials (Falkman, 2001).

As recently as 40 years ago, packaging primarily referred to fiberboard, wood, metal, glass, and other natural or organic packaging materials such as real horse hair. Today, science and technology have changed things so drastically that packaging engineering is primarily a discipline you will find as a track or specialty or minor within undergraduate plastics engineering programs. Nonetheless, fiberboard is the most universally used packaging material once one gets beyond the retail package; and paper, even if coated or treated, is still the most frequently used retail packaging material.

Products of all sorts now travel over greater distances, requiring more handling and making them subject to more severe conditions for longer periods of time during the transportation distribution cycle. Packaging—even at the lowest level, the retail package—has undergone huge changes yet it still represents a giant direct cost for the producers of goods and represents a growing challenge in terms of solid waste, landfills, and related environmental considerations. If one looks around today, one also sees a change in the way information is provided on packaging at all levels. For a very long time, until late in the 20th century, in fact, printing inks were primarily petroleum based and were flammable liquids; thus, the inks themselves were treated as Class 3 regulated materials. Today, we see a move towards soy-based inks; this is just one example of the changes that are occurring. In the past, we used words and symbols within the supply chain, especially for transportation, to convey information. Today, we use 2-D and 3-D bar coding; IR readers on our phones; and RFID chips on individual items, shipping cartons, and shipping containers. RFID chips are used to provide more information, in real time, to obtain in transit visibility, or to get or exchange other real time information at or near the point of purchase.

4.3 Risk Threat and Vulnerability

4.3.1 Overview

Table 4.7 and Figure 4.3 help underscore the point that hazardous material exists all around us, and that in general very few people have any awareness of, much less an understanding of, the hazards that surround us at work, at home, and in between. The public at large, the business community in general—especially the retail segment—and government at all levels from global to national to state and tribal to local, do not understand, or have uniform methods for, assessing and measuring the risks. For that reason, it is impossible to address the risk from any universally recognized and accepted risk threat and vulnerability assessment consensus or statutory standard, and then take broadly accepted corrective actions. Many everyday items are hazardous materials or can easily combine with other materials to create hazardous materials. Chlorine bleach and ammonia, combined, produce highly deadly chlorine gas. Hydrogen peroxide is a commonly used antiseptic in a very dilute solution. What very few know or understand is that 100% hydrogen peroxide (CLASS 5.1 OXIDIZER and CORROSIVE) must be chemically stabilized before it may be moved.

4.3.2 Risk Threat and Vulnerability

There are separate courses and texts for risk threat and vulnerability, liability issues, and mitigation, and minimization and prevention strategies, but we are going to take a fairly broad view here to impart ideas and raise awareness. Nonetheless, we will look at specific examples and opportunities as part of this unit to demonstrate opportunities and ideas. Before getting into the meat of the material, we do need to discuss these terms and how they fit into other disciplines and programs of study. Today, in the post-9/11 world, risk and vulnerability assessment is too often thought of as part of homeland security. That is a very shortsighted and inherently incorrect view. The terms threat, vulnerability, and risk are three regularly mixed up and not fully understood terms. One framework to help understand them is illustrated in Figure 4.4.

Another way to look at this in a slightly abbreviated way is to use a form of a slightly more focused approach. “Vulnerability analysis involves an understanding of the nature of the system, and of the *likelihood* of threats/impacts. Threats are examined on the basis of their likelihood.

Table 4.7 General Use and Household Hazard Identification Matrix

Material	*Class*	*Example*
Calcium hypochlorite	Oxidizer	Algicide, swimming pools
Sodium hydroxide	Corrosive	Detergent and soap (TLV)
Sodium hypochlorite	Corrosive	Swimming pools, laundry
Nail polish, polish remover, perfume/cologne, most stains, and some paints	Flammable liquid	Self-explanatory
Liquid-flavoring extracts	Flammable liquid	Vanilla extract, lemon extract
Herbicides, pesticides, fungicides	Toxic	All those gardening chemicals
Fire extinguishers and hair spray	Compressed gas	Self-explanatory
Correction fluid	Toxic	Office
Toner	Flammable	Office
Ink pen cartridge	Flammable	Office
Printing inks (older formulations)	Flammable	Local sign and printing companies
Brakes	Asbestos	Cars and trucks
Smoke detectors	Radioactive	Home and office
Fluorescent bulbs and starters	Toxic	Everywhere (school)
Older fluorescent ballast (PCBs)	Class (carcinogenic)	Self-explanatory
Batteries	Corrosive and explosive	Everywhere
Markers	Flammable	Classroom
PC CRTs media hard drives	Toxic	Chromium, beryllium phosphors
Oxygen generators (all types)	Oxidizer	Aircraft and personal medical use
Airbags	Explosive and corrosive	Cars and trucks
Lithium-ion batteries	Flammable solid	Consumer electronics, toys, office devices, aircraft black boxes, and much more

Impacts are evaluated on the basis of their scope" (Disaster Center, 2002). That is a simplified approach that inherently recognizes the material above.

The larger concept would be better characterized as a primary life process and requirement. If one looks at various disciplines and programs, one will find terms such as *security management,*

Awareness includes recognition of hazardous material, recognition of releases of hazardous material, and the ability to take appropriate safety and response actions. A brief summary of some of these concepts follows:	
a. Recognizing releases is based primarily on the basic senses and an understanding of employer-installed monitoring systems. (Remember that all warning systems depend on sensory stimulation.)	Sight, sound, smell, touch, and taste. You can see liquid leaks or vapors. You can hear gas leaks, fire, and boiling. You can smell many things. You can feel or sense rising or dropping temperatures or anything falling on you. You may get a funny taste in your mouth when working in an area.
b. Ability to take appropriate response has three parts.	Understand the hazard. Know the response. Know what safety devices, including alarms, fire-fighting materials, breathing apparatus, protective clothing, spill containment, and isolation mechanisms are available and where they are located.

Figure 4.3 Practical awareness.

An asset is what we set out to protect (that includes you as an individual).

Threat—Anything that can exploit vulnerability, intentionally or accidentally, and obtain, damage, or destroy an asset.

A threat is what we're trying to protect against.

Vulnerability—Weaknesses or gaps that can be exploited by threats to gain access to an asset or inherent weaknesses in the as-built environment around the asset.

A vulnerability is a weakness or gap related to the asset(s).

Risk—The potential for loss, damage, or destruction of an asset or organizations, or damage to external entities as a result of a threat exploiting vulnerability.

Risk is the intersection of assets, threats, and vulnerabilities.

Why is it important to understand the differences among these terms? If you don't understand the difference, you'll never understand the true risk to assets. When conducting a risk assessment, the formula used to determine risk should be: **A × T × V = R.**

Figure 4.4 One framework for threat risk and vulnerability. (**Modified from the Threat Analysis Group, *Threat, Vulnerability, Risk—Commonly Mixed Up Terms,* 2010. Available from http://www.threatanalysis.com/blog/?p=43.**)

business continuity, continuity of operations, and *continuity of government*—all evolving from the elements, based upon risk and vulnerability assessment and liability issues.

For our discussion, it is extremely important to separate out the two terms "risk" and "vulnerability." They are not the same, yet many people continue to use them interchangeably, failing to recognize that they represent two entirely separate ideas. Everything we do contains some risk—a point that is missed by many. When one undertakes a risk assessment or creates a risk inventory, one is examining possible outcomes if something goes wrong. That is still a very broad statement, but that is the way we will use the term here. Earlier in the book we talked about a serious failure in a water tank in Rochester, New Hampshire, and we mentioned the Boston Molasses Disaster. Let us look at some of the ideas and concepts related to risk vulnerability and liability by using these as examples. The threat here is created by the storage of very large quantities of liquids (one extremely viscous molasses and the other, very fluid water) in above-ground tanks, so the risk is created by the act of storing these large quantities of liquids. Looking further, the molasses was in a cast iron tank while the water was a steel tank. A separate observation, as we proceed to discuss the issues here: the opportunity existed for either of these large quantities to be stored below ground rather than above ground. Now look at some of the ways we can apply analysis and science to determine how to assess the "risk."

1. The inherent or primary risk is created by the storage of the great quantity of liquid regardless of its composition or chemical makeup.
2. The second risk relates to the two materials: water holds very little capacity to damage beyond its sheer volume and weight. Molasses is a more complex liquid that has additional and/or different characteristics and therefore poses additional risks. Regardless of intent or normal conditions, molasses is extremely viscous, and, molasses, which is an organic sugar, can ferment. This process generates gas and builds pressure in the container. This process also produces alcohol. While fermentation may be part of, or the purpose of, the production cycle, the possibility of early fermentation in the tank, preproduction, must be taken into consideration. That requires additional monitoring and/or remediation processes, which might include neutralization, pH adjustments, pressure relief or cooling, and possible treatment of the vapors released, as well as the recognition of the possibility of explosion or total destruction of the holding tank. None of those exist for the water.
3. The next issue is the physical configuration of the tank. In Boston 1919, we are talking about a tank on top of the building; in Rochester 2011, we are talking about a tank sitting on the ground and technically just below ground level.
4. The next important element of risk in this case is the as-built environment around the two separate tanks. In Boston the tank was located in a heavily populated urban area with distinct peaks in population during the day and decreases in population during the evening. In Rochester the tank is located relatively in a rural area, well outside of this small city's dense urban core, and the daytime/nighttime population variations would have been smaller.
5. The last risk element is control or security, although that did not have any impact on either incident. In Boston the tank was within the physical compounds of the owning company's facility/property so there was inherent physical security. In Rochester the tank was an isolated structure in a mixed residential commercial low-density zone.

That identifies the "risks" associated with these two tanks. Next let us talk about possible vulnerabilities.

1. The molasses tank was within the physical perimeter of the owning company, but it was within a heavily populated area. This means it was vulnerable to fire and other accidents, including the possibility of a vehicular or elevated railway extreme condition. In addition, because of the nature of the product stored in the tank, the tank was vulnerable to catastrophic destruction of the tank itself due to explosion or if you prefer pressure buildup in the tank.
2. The water tank is relatively near a major numbered state route that receives considerable traffic and is a major artery for the local economy but otherwise it is not in a hostile or heavily populated area. Therefore, a partial failure of the tank itself becomes the vulnerability.

The next element to examine is threats. In this case, we have to keep in mind the very different times these two events took place. Knowing what we know today, there is a definite unique threat that needed to be taken into consideration in 1919 and an entirely different threat that must be considered today. I would suggest that there is only one unique threat that could or should be attached to each incident. Given the specific time frame of the Boston molasses disaster, the threat would have been specifically for theft and use in illegal additional activities. Molasses can also be fermented to produce rum and ethyl alcohol, the active ingredient in other alcoholic beverages and a key component in the manufacturing of munitions at the time (Puleo, 2004). Given the inherent importance, not value, of the municipal water tank, the threat in 2012 would be a terrorist attempt to sabotage a municipal water supply.

Let's think about this and talk about it a little more. In the Boston molasses disaster the fact that the tank was within the perimeter of the facility would indicate that additional physical security might have been appropriate if the threat of theft by bootleggers is considered real. In essence, although this was a legitimate threat, the tank was not necessarily vulnerable or, if one was to claim there was a certain amount of vulnerability, that problem could be corrected by the addition of physical security (a fence and armed guards). On the other hand, the vulnerability of the Rochester municipal water tank would not be easily addressed, although its location in a populated area and in plain sight makes it a harder target for terrorists than one might assume, given the heightened awareness in the country today.

4.4 Liability Issues

That brings us to the next issue of liability. Since we are using the Boston molasses incident as an avatar for commercial or industrial private sector operations, we are looking at liability in terms of current OSHA and EPA requirements and in the context of federal and international legislation affecting the environment. Given the storage tanks close proximity to the waterfront, there might very well be additional Coast Guard or public health liabilities because of that proximity to the water.

There are a very large number of inherent, if unstated and indirect, liabilities associated with the molasses tank in a heavily populated area. There are liabilities related to the storage of such large quantities of any liquid—liabilities attached to personnel safety and maintaining the tank itself. There are also a full range of OSHA and EPA considerations, given that this particular product can ferment and is subject to large pressure changes due to chemical activity within the material itself when held in such large quantities. There is an inherent liability because of the large population surrounding the facility, and there would be environmental liabilities attached to the processing of the material. On the other hand, for its part, the City of Rochester has to deal with either the loss of the water resource or a partially catastrophic large-scale release of the water where mechanical action of the water itself is likely the only threat to the as-built structure and the population surrounding the tank. Let's touch on just one more remedy to consider that might apply to either or both situations: below ground storage rather than above ground. For the City of Rochester, with a population of slightly less than 30,000 inhabitants according to the 2010 census, the acquisition of the land necessary and the preparation of the site would add considerable cost with no intrinsic advantage in an area where land is relatively cheap.

In a perfect world, all regulatory compliance costs could be clearly identified and effectively allocated to product; thus, they could be passed on to consumers. Even then, consumers would have to be willing to pay for a healthier environment and safer workplaces. It would be a simple process to eliminate hazards within an organization's operations and thus avoid liability issues, high insurance rates, and public dissatisfaction with the way the organization does business. It would not eliminate excessive oversight and overregulation on the part of local, state, tribal, and federal governments, but it would at least allow for cost sharing between producer and consumer. Dream on!

As we point out in our discussion of plating processes, there are many choices to make even in the selection of something as simple as a screw or other fastener. A life cycle approach requires that every hazardous material used within the producing organization be analyzed. In some cases, it may be possible to replace the material with another material that either is less toxic or is not toxic. It may also be possible to replace a hazardous process with the use of a nonhazardous process. It may also be possible to outsource processes that either require use of hazardous materials or processes that are so inherently dangerous that they would be better performed by a company specializing in the process. One must also look at processes and functions in another way as well. The use of materials and products that ordinarily are considered not hazardous may generate regulated waste streams and/or specifically taxed environmental impact waste streams. Every material process and procedure that uses or generates hazardous materials or wastes needs to be identified and examined to see what, if any alternatives exist. This also needs to be extended to what previously were traditionally considered incoming logistics streams smaller component parts and assemblies. If their production carries a greater than normal hazardous materials component or environmental component, that, too, must be factored into the design build and purchase decisions. It is no longer adequate for the producer of goods to merely look at form fit and function.

As part of this process, concepts that are captured in Bierma and Waterstraat's book *Chemical Management* need to be applied. The full, and substantial, EHS cost or hidden cost for the use of hazardous chemical products and processes can dramatically increase direct and indirect costs while adding considerable layers of risk and liability. We have not mentioned it before, but the insurance industry recognizes all those risks and liabilities, and therefore your insurance rates get driven up when you use hazardous materials and/or processes or purchase materials that represent a hazard that you might eventually be held liable for. One very important fact that needs to be reintroduced at this point is the ownership of hazardous wastes. In essence, if you are in a position where you generate waste at the point of production, or you are under a growing body of regulation that assigns costs back to the producer for the eventual disposal of products by retail users, you essentially own the waste unless or until you can prove it has been converted to a nonhazardous form. That consumer disposal stream is just recently (2010) coming under wider regulation and having costs assigned back to the manufacturer; therefore, many businesses may not yet be aware of it and those costs not likely to be in many business plans and risk assessments yet. That has been demonstrated many times in the courts. Today, especially in view of changing regulations for products that would be described primarily as consumer or retail products, that is likely to hold true for the original producer of the goods rather than any intermediaries in the distribution chain or the retail consumer/user.

In the United States we are seeing growing statutory requirements that directly affect consumers and households in regard to the disposal of hazardous materials. To date, there are no effective ways to track or enforce those regulations. The awareness education component, which is critical, is sorely lacking in the government's approach to those hazardous waste streams. Remember, there is already legislation in many nations attaching recovery and recycling costs to products as they leave the manufacturing facility. Chemical streams—and that includes everything from janitorial supplies, to boiler water treatment, to routine lubricants and office supplies to household cleaners, fertilizers, and pesticides—represent areas where hazardous materials can be found and where hazardous waste is generated.

There are essentially two distinct sets of streams that the manufacturing or producing organization can be, and are being, held responsible for: (a) solid, liquid, or gaseous streams, that are generated while producing goods and (b) the waste streams, which are very similar to nonpoint pollution, that result from billions of retail users disposing of hazardous materials.

4.5 Mitigation, Minimization, and Prevention Strategies

In many cases, there are no realistic and suitable substitutions for the processes and/or the products we are talking about. However, as technology continues to evolve and materials science advances, we may in fact see the elimination of many such processes and products. In the meantime, it is wise to examine alternatives that mitigate, minimize, or prevent. The range of problems and solutions is vast, but there are very large number of options and alternatives available right now that range from something as simple as crushing and recycling plastic, wood, and metal containers to turning wastes oils of all sorts and petrochemical based solids and liquids into feedstocks for furnaces.

Here are a number of examples of areas that have already been proven or are being looked at for recovery exploration and expansion:

Fiberboard boxes. Should fiberboard boxes be recycled into new fiberboard products, burned as fuel without any additional processing, processed, and then burned as fuel, *or* allowed to go into solid waste facilities? If disposed of in solid waste facilities they will decompose and generate methane gas. There are numerous examples of high-quality commercial recovery of methane gas from landfills. "...One project provides the UNH campus with landfill gas from Waste Management's Turnkey Recycling and Environmental Enterprise facility in Rochester, New Hampshire, approximately 13 miles away .... The program currently provides the campus with up to 85% of all its energy needs" (UB Custom Publishing Group, 2011).

If you are looking for an answer to that question in the text you will not find it; there are too many variables at play here, and the technology is changing more rapidly than the author can easily track. As an example, if you are expending energy to process the fiberboard before you are using it as a supplemental fuel, will the fiberboard yield enough BTUs to offset the additional cost of processing it before it is burned? As primarily a business text rather than a scientific treatise, we will not attempt to gather the analytical data to support any answer to that question.

Transportation packaging materials. It is now understood that untreated wood is an inherently unsuitable media for packing or protecting goods over long or short distances, especially when such a move involves significantly different ecosystems. Wood can play host to various living orgasms such as arachnids nematodes, and, it can carry bacteria blights and spores. Moving agricultural, aquacultural, finished, and/or forestry goods, as little as one county in the United States can lead to significant infestations and huge economic losses. Globally, there are now two acceptable—or perhaps mandatory would be the better term—treatments for wooden packaging: heat treatment or methyl bromide treatment. The current standard is ISPM15; you can find this through USDA APHIS or going directly to the international plant protection convention website. In recent years, there has been one scare about Tylenol being contaminated by engineered wood shipping with transportation packaging materials. Since methyl bromide is designed specifically to kill living things, it should not be a surprise, although it may not be obvious, initially, that it can kill humans as well. The author is not suggesting that methyl bromide is a bad treatment but only pointing out that as time goes on, we recognize more risks from the increased use of all chemicals. Thus, if there is a way to use something that has not been subject to any form of chemical treatment, you are probably better off.

Construction treated wood. This is used in any number of applications in residential construction. Most people do recognize treated lumber that goes outside and into the ground. What people may not recognize is that the treatment is usually chromated copper arsenate (CCA). That is what gives it greenish tinge. CCA is a pesticide. That means it kills living things! Many of the engineered wood products now used in homes have been subjected to varying chemical treatments and/or use various adhesive. Because of that you will notice, if you take the time to look, that many such products are marked one way or the other to inform you that they are not to be burned and absolutely never burned inside of a home. Again, the author is not suggesting that such products need to be reengineered or that they should not be used, rather, that they all represent inherent hazards and as such also represent opportunities to reduce corporate as well as individual risk.

Going back to fiberboard packagings we are now seeing movement away from petroleum-based markings to soy-based markings and the use of different adhesive in the manufacture of those containers. The initial purchaser or user of packaging materials needs to ask these questions. Petroleum-based inks actually make more sense if the life cycle plan for the packaging assumes its use as feedstock for furnaces. Soy-based inks might make more sense if the packaging is designed to go into landfills, which mean that it will eventually lead to the generation of methane.

Cases, shells, or covers for home appliances, consumer electronics, and large toys. These all offer another unique opportunity for more basic material recycling through grinding, shredding, and reprocessing; this is still fairly new as a best practice or design practice, but the fact remains that there is inherent value in scrap cases from no-longer-being-used consumer electronics, including computers, monitors, printers, TV sound systems, home appliances, larger riding toys, lawnmowers, and many other products. Neither metal nor plastic nor composite provides an automatic advantage in terms of life cycle management and recycling, but mixing two or three into the outer "form" of a product creates problems in terms of separation and energy usage to convert such materials into usable streams. In most cases, that means grinding, shredding, or melting for reuse of the materials. Decomposition in other cases may mean grinding or pelletizing for fuel, and, in other cases, it may mean grinding or shredding for reuse and as additives to new materials or for completely different uses. While there are few large items currently encased in glass, larger glass containers, along with aluminum, tin, and plastic, also fall into the category of easily recyclable materials of relatively high value. The dual issues of increased regulatory oversight of and new or preferred disposition processes and new uses for previously unrecoverable wastes may drive both the choice of material, the mixture of different materials, and the method in which visual information is made on or as part of the "case" or outermost surface of both products and packagings.

Rotating machinery. For many rotating electrical devices, there are also issues related to design and application. Areas that need to be studied include the nature of the "brush" material used to conduct electricity from the static structure to the rotating element; the need for, and the types of, bearings; treatments or coatings to reduce friction: choice of lubricants or designs to eliminate the need for lubricants; and selection of gaskets, O-rings, and packing materials. A side issue for large applications is heat generation, cooling, and/or heat recovery to improve performance, enhance reliability, extend service life, and reduce energy usage/costs. If you look at 2012 literature on ocean liner design, you will see that main shafts are moving towards seawater lubrication rather than petroleum or synthetic "oils." Given the growing concern for, and awareness of, ship board bilge and waste products, and the new legal requirements requiring ballast water exchange (BWE) and bilge treatments onshore, the inherent advantages are quite large. If one can convert from synthetic petrochemical lubricants to a forced seawater lubrication system, a number of cascading impacts occur. The need for processing companies to produce lubricants is reduced; the need to ship those lubricants to the liner is eliminated; the need to design space for, and reserve weight for, the lubricants is eliminated; the need for treatment of the waste product is eliminated; and the inherent possibility for pollution is reduced.

Batteries. If one follows the literature on batteries, it also becomes apparent that there are inherent risks with lithium-ion batteries, although design improvements and changes in technology are reducing, not eliminating, such risks. Given the present trend in regulations, driven by a growing body of knowledge, it is reasonable to assume that during our lifetimes lithium-ion batteries will have to be treated as hazardous waste and inherently hazardous in households, as well as in the commercial work space. Given that scenario, it is not unrealistic to assume that there will come a point in time at which it will be able to classify different types of batteries as being "green" or preferable choices for different fairly broad applications. We are not there yet.

Lighting. This area was mentioned earlier but it is worth reviewing lighting—commercial, governmental, and private/residential—to review where the costs and risks exist in the near term and remind you that it is necessary to stay aware of changes in technology, scientific research, and regulatory responses. All forms of fluorescent lighting have inherent risks such as the use of mercury; older fluorescent lighting fixtures may also be using ballasts containing PCBs. All forms of incandescent lighting are highly inefficient as the basis of incandescent lighting is high heat, which in turn generates light, therefore the maximum energy used to produce light is wasted as heat energy. Mercury vapor and halogen lighting definitely have limitations. Current technology seems to be driving towards LEDs, but that is not to say that sometime in the near future some as yet undiscovered issue will reduce the inherent advantage of LEDs. The initial cost is no longer, and has not been for some time, the most important issue, although popular literature has led many to believe that to be true.

Building and construction materials. For convenience, and in an attempt to bring this home to readers as individuals rather than as workers, let us talk very briefly about building construction materials and techniques. We are already seeing large-scale replacement of plywood exterior sheeting with oriented strand boards (OSBs) and other manufactured materials. In recent years, we have also seen plastic wrapping become standard process in building construction. Technology and science are leading to more and more applications of geothermal and solar systems to reduce home energy costs. We are also seeing increased use of steel two-by-fours replacing traditional wooden two-by-fours. What you may not be aware of are the realistic application of straw as a building material within specific geographic areas; such structures exist as far north as northern New Hampshire. There are some obvious drawbacks to hay bale construction but like every product, there are pros and cons. The biggest drawback to increased use of such materials is overregulation, as distinct from effective science-based regulation, in regard to building codes zoning and enforcement. This one area alone is worth a little extra time on your part in order to understand what "innovation" really means and what "natural" really means. If one understands and accepts both of these core concepts, one will then be able to grasp the problems generated by an overly restrictive and intrusive federal framework.

Landscaping. This is included here for two reasons: Green lawns in general and improperly selected ornamental plantings create multiple problems. Lawns may be aesthetically pleasing but they are in fact "unnatural." The nature and quantity of fertilizers used to maintain "green" lawns in the United States is perhaps the largest non-point-source form of water pollution, but very few planning boards and homeowners recognize or understand that fact. Lawns also consume unreasonably large quantities of water, and water consumption is one of many issues related to improper selection of ornamental plantings. Some such plantings are invasive or are hard to control, while others can provide invaluable filtration, remediation, and erosion control. Proper selection of plantings and proper retention of existing trees and shrubbery on "to be developed" industrial, commercial, and housing tracts can all reduce heating and cooling costs and, to some degree, and offset the impact of large increases in impervious surfaces that create significant sheeting and runoff problems.

In this and the preceding two sections, we have covered a large volume of material that spans many different ideas and areas. The bottom line is that science and technology offer us opportunities to vastly reduce our consumption of finite resources and reduce the impact of all sorts of environment impacting development/expansion. In the past, the world has been able to ignore the concept of finite resources, but population growth is making that a larger challenge. Couple that with the growing body of knowledge that demonstrates the many negatives of ill conceived and implements industrialization and both overregulation and lack of scientifically based and technically sound regulation, and one better understands why it is necessary to act regardless of the compliance requirements in the existing regulations.

4.6 Stakeholder Identification

This is a critical element and an often misunderstood, ignored, and misapplied concept with respect to hazardous materials, hazardous wastes and life cycle management. It would be quite easy to make the business case that everyone is a stakeholder, but that is neither realistic nor effective. Costs related to life cycle management of hazardous material can be categorized as including all of the following: economic impact of cost; time costs within the entire system; environmental health and safety costs; public perception and marketing costs, including company image in the community at large; and, regulatory oversight costs. If one accepts these, it becomes possible to identify larger groups, and in some cases subgroups, of stakeholders who need to be engaged from the earliest stages of product concept and the development process. The owners of a company, whether public or private, qualify as the first major stakeholder group. The second two groups, all directly impacted, are the top management followed by all the other employees. While all of them are employees in the largest sense, they do represent two distinct subgroups within a larger group.

All an organization's suppliers and external service providers are stakeholders.

From the regulatory and governmental perspective, environmental, occupational, transportation, and, health-regulating authorities and enforcement organizations represent a critical group of stakeholders. Some could argue that this automatically includes the entire first responder community: law enforcement, fire, dispatch, and EMS. They are, by their very nature, the first line of defense, primarily user-reactive response elements. They most certainly are stakeholders those involved in the regulatory and permitting process are more important stakeholders. In the end, the first response community is a community one hopes never to have to see at their facility. Within a company, maintenance personnel for production machinery, as well as those charged with facility maintenance and general upkeep, and shipping and receiving personnel are a group that are essentially on the front lines, and it makes sense to identify them separately and specifically within any organization. They are the ones who really understand what is going on as distinct from how the product is perceived or accepted by the public the producing organization serves.

Stakeholder involvement needs to extend beyond classical and traditional product teams, which are often created in highly controlled environments where management cultures prevail. From fairly early in the process of concept, not necessarily product development, external eyes, especially all supply chain partners and projected users, provide critical insights for successful product design and development, which in turn leads to effective life cycle management of the product.

References

Alderton, P. (2008). *Port operations*. London: Informa.

Bierma, T., & Waterstraat, F. (2000). *Chemical management*. New York: John Wiley & Sons.

Eartheasy (2012). *LED light bulbs: Comparison charts*. Retrieved from http://eartheasy.com/live_led_bulbs_comparison.html.

Falkman, M. A. (2001, August 1). Packaging function, Structure defined. *Packaging Digest.*

Griffin, R. D. (2009). *Principles of hazardous materials management*. Boca Raton, FL: Taylor & Francis.

Ivanco, M., Waher, K., & Wardrop, B. W. (2009). *Impact of conversion to compact fluorescent lighting, and other energy efficient devices, on greenhouse gas emissions*. Retrieved from http://www.intechopen.com/source/pdfs/11764/InTech-Impact_of_conversion_to_compact_fluorescent_lighting_and_other_energy_efficient_devices_on_greenhouse_gas_emissions.pdf.

Maloney, J. C. (2003). *The history and significance of military packaging*. Fort Belvoir, VA: Department of the Army.

McKinlay, A. (1995). *Need a new direction for your transport packaging? Do a 180!* Chicago, IL: IOPP.

McKinlay, A. (1998). *Transport packaging*. Chicago: CRC Press.

Moll, E. (2012). *Energy-efficient bulbs: halogen vs. fluorescent vs. incandescent*. Retrieved from file:///C:/Users/User/Desktop/Electronic%20Documents/energy-efficient-bulbs-halogen-vs-fluorescent-vs-incandescent-3228.html.

Puleo, S. (2004). *Dark tide: The great Boston Molasses Flood of 1919*. Boston, MA: Beacon Press.

Ramsland, T. (2000). *Packaging design: a practitioners manual*. Geneva: International Trade Centre.

Threat Analysis Group. (2010). *Threat, vulnerability, risk—commonly mixed up terms*. http://www.threatanalysis.com/blog/?p=43.

UB Custom Publishing Group. (2011, October). Landfill gas powers higher education at University of New Hampshire. *University Business*. Retrieved from http://www.universitybusiness.com/article/landfill-gas-powers-higher-education-university-new-hampshire.

Chapter 5

National, Tribal, State, and Local Government Considerations and Challenges

5.1 A Framework for Further Discussion

5.1.1 Jurisdiction Authority Responsibility

Any system designed to segregate areas is by definition arbitrary. For the purposes of this chapter, the primary categories we will use include the following:

a. **Internal government operations.** This topic deals with the administrative functions necessary to deliver government-provided goods and services. Government at every level and however small will definitely have a range of these activities.
b. **Business and industrial operations within the organization.** The majority of governmental entities will have at least some level of business and industrial operations, even if that is no more than the operation of a fleet of vehicles.
c. **Commercial operations within the jurisdiction.** This would include goods providers, service providers, NGOs, and all the utilities and transportation sector businesses, and of course the entire hospitality industry and the retail outlets—as well as medical and dental facilities and services to name just a few—that exist in the community.
d. **Private and consumer activities within the jurisdiction.** This refers specifically to private individuals in their role as residents of the community who consume the goods and services provided by and through the local government. This includes the employee role, the employer role, the commuter role, and the dwelling resident role or visitor role for those who are enjoying the goods and activities available in the community.

5.1.2 Overview

This chapter addresses all levels and all facets of government. It is based primarily on the American model that conceptually should have little impact on most of the material. Government is primarily in the business of providing services rather than goods. Clearly, there are exceptions or unique circumstances that drive government into the production of goods rather than services. Here are two examples: (1) the US Department of Defense, DOD, contracts for and "makes" its own goods; it also performs major industrial operations on large weapon systems and everything from ships, planes, trains, and tanks to printing books, thus taking on most, but not all, the characteristics of a private entity designing building and delivering materials; and (2) prisons, typically state prisons, where inmates are producing real or tangible products. We may refer to these throughout this chapter, but we are not going to address the issues related to industrial commercial operations. For that, one must go back to the information in the previous chapter addressing private enterprise. There are also a full range of services that require industrial activities. For that reason, a lot of the material from the private sector chapter will be repeated here as well. The key point is that the bulk of government operations is focused on providing services in the administrative sense, so those are the issues that will be treated in greatest depth in this chapter.

Figure 5.1 represents all levels and forms of government. However, the figure might be redrawn to show the industrial function block as a much smaller-sized block for all levels below the national level. The US federal government, primarily because of the activities in the Department of Defense and Department of Homeland Security and the General Services Administration (GSA), engages in significant industrial activities that warrant the same approach as a goods-producing enterprise. In contrast, lower levels of government certainly embrace certain industrial activities, but they represent a much smaller portion of the governmental entity's base operations. For that reason, a lot of the information in this block will not be applicable to the vast majority of governmental units using this text.

Figure 5.2 is a better representation of the "function" of government and the very large primary consumers of its goods and services. Government exists to serve its constituents, and the major constituencies consist of two broad groups depicted here as private citizens, and business and industry. The term *business* as used here applies to those that deliver goods and services as distinct from those who produce tangible goods. It is a conceptual convenience, not an absolute definition. Technically, agriculture could be shown as yet another of the constituents, but many smaller agricultural operations contain elements of both of the major segments identified above. In Figure 5.2

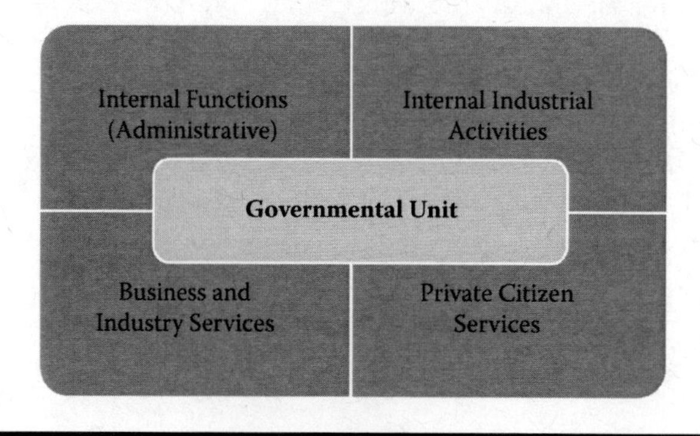

Figure 5.1. Functional view of a business enterprise.

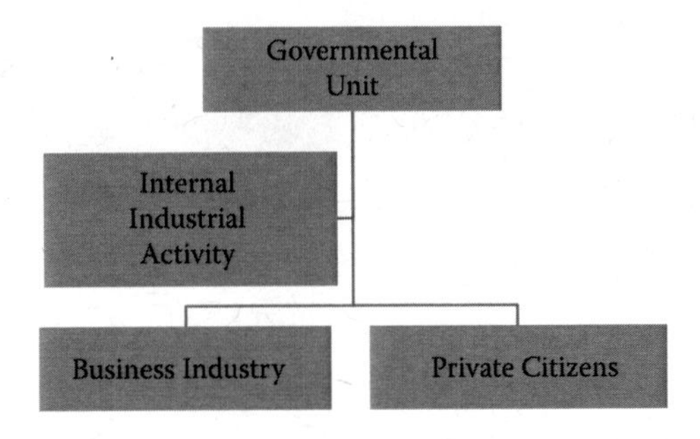

Figure 5.2 A flow diagram for government operations.

two constituencies each must be viewed from two separate perspectives. One is the provision of services. While much of this may be administrative, this portion also contains a separate element of infrastructure services and public safety services. The other perspective is oversight compliance and regulatory functions. It is primarily the administration of those regulatory or oversight functions that involve areas directly related to hazardous materials and life cycle management.

In contrast to Chapter 4, some areas that were covered at great length and great detail will be ignored or minimized. This goes back to the concept of recognizing government as being in existence to provide services to its constituents. While there are very significant differences in the "services" provided by state, local, and tribal governments, the material in Table 5.1 shows that all of the following functions can be found at many levels of and within a number of different-sized governmental entities: law enforcement, fire protection and emergency services, public health services, highways and bridges, water and sewage, gas and electricity, mass transportation services, and recreational services and facilities.

5.2 Recognizing Defined in Undefined Hazardous Materials, Hazardous Material Streams, and Waste Streams

In early 2012, a number of recycling projects, not necessarily hazardous materials recycling projects, have been highlighted in various industry publications. Below are excerpts/extracts highlighting some of these more unique activities. As a group, they should give government at the local, state, and national level ideas about how to reduce all forms of waste, not just hazardous waste:

1. The Connecticut DOT is working in a partnership with seven railroads to redistribute materials from a 5 mile track replacement project. The seven railroads, including the Providence & Worcester Railroad and the Housatonic Railroad Company and four rail museums, including the Shoreline Trolley Museum and the Central Connecticut Railroad Museum, will be able to step up their efforts to rebuild and strengthen rail infrastructure on branch lines throughout the state. What most people might not recognize is that the bulk of older railroad ties are creosote-soaked timbers. For disposal purposes these are hazardous waste stream so recycling has triple benefits
2. Arizona and Nevada add tons of recycled rubber tires into road resurfacing material. In this case, there are specific roadway improvements that include wide-area safer rides; reductions in raw materials needed; and considerable reduction of solid waste and waste products that represent multiple safety and environmental hazards. There have been some very serious

Table 5.1 The Governmental Construct

Internal functions	**Exposure risk opportunity**
Administrative/office Facility management and maintenance Janitorial Support operations Purchasing of raw materials Purchasing of parts and supplies & consumables Product packaging Transportation packaging Distribution functions, including storage warehousing transportation	Management of delivery models, understanding of best business practices and delivery models and modalities Feedback from frontline personnel and the public* Office supply quantities and materials* Lighting, heating, and cooling* Utilities and space* Cleaning supplies and disposal methods*
Industrial functions	**Exposure risk opportunity**
Purchasing of raw materials Purchasing of parts and supplies & consumables Product packaging Transportation packaging Distribution functions, including storage warehousing transportation	Management of design: understanding of practices and materials to be avoided* Product selection criteria going beyond pricing Chemical processes; chemical selection and chemical recovery; treatment* Material usage and selection Transportation packaging* Material handling equipment* Transportation methods* Transportation providers*
Business & industry activity services	**Exposure risk opportunity**
Regulation and compliance activities Service activities	Understanding of practices and materials to be avoided in the business community* Understanding and control of chemicals used across the entire commercial sector of the community General and specific recycling programs Education programs and incentives
Private citizen activity services	**Exposure risk opportunity**
Regulation and compliance activities Service activities	Understanding of practices and materials to be avoided in the home environment Effective and appropriate regulations and ordinances Understanding and control of chemicals used and sold in retail establishments and used by private citizens Effective and accessible recycling facilities Effective and meaningful outreach and education programs

* Material appears in Chapters 4 and 6 as well.

rubber tire dump fires. Perhaps the most relevant one is the tire fire underneath Interstate 95 in Philadelphia that closed one of the most important commercial highway corridors in the United States. Rubber tires themselves do not decompose in solid waste landfills and when disposed of indiscriminately tires provide breeding grounds for mosquitoes.

3. Kentucky contracts for about 700 construction projects each year. Contractors had to purchase bid proposal packets containing upwards to 400 pages each for every project in which they were interested. In addition, bids had to be submitted on paper and delivered in person. Beginning in April 2010, bidders were required to register online, eliminating mandatory trips to the KTC offices and the need to print an estimated 200,000 pages of bid-related documents each year. The organization announced this as a huge reduction in paper. But what it failed to recognize or address was the huge reduction in printing costs, from a hazardous material standpoint that it has to do with the inks and/or toners that would otherwise have been used—such products do become regulated waste or require special handling for recycling.

In addition to these two illustrative cases that have direct hazardous material hazardous waste impacts, there were a number of other noteworthy state DOT efforts highlighted that demonstrate the many ways governmental and quasi-governmental entities can improve efficiency, lower costs, and reduce wastes. These include the following:

- The Louisiana Department of Transportation and Development pilot program used a new LED solar-powered airport lighting system that is expected to reduce the airport's taxiway energy consumption by more than 90%. The solar panels charge a bank of batteries that can power the taxiway lighting system up to 14 days with little or no sunlight.
- The Oklahoma Department of Transportation (ODOT) took an I-40 bridge out of service. Instead of tearing down and scrapping the 47-year-old structure, ODOT decided to recycle the bridge, awarding a $10 million contract to deconstruct it. A majority of the bridge's 1900 beams are expected to pass inspection and be made available at no cost to county governments for use in bridge projects statewide. ODOT estimates that the recycled beams could help counties construct more than 300 bridges, saving Oklahoma taxpayers millions of dollars.
- The North Carolina Department of Transportation (NCDOT) now allows for the recycling of residential roof shingles in the production of asphalt pavement. About 260,000 tons of used shingles went into landfills in 2011. If this many shingles were diverted from landfills, the annual cost savings would be about $32 million.
- The Florida Department of Transportation (FDOT) estimates that by July of 2012, an array of more than 400 solar panels will provide 70% of the electricity used at its new turnpike retail facility at the Turkey Lake Service Plaza.
- In 2011, Michigan Department of Transportation (MDOT) began using a combination of wind and solar photovoltaic energy systems to power three rest areas and two welcome centers. A 100-kilowatt solar array was also installed at a rest area near Grand Rapids, to generate energy for freeway interchange lighting and to be used for the entire metropolitan area. MDOT also has installed geothermal heat pump systems at 10 rest areas, and they will be included in future rest area projects where feasible. MDOT is also using LED light fixtures to illuminate both the interior and exterior of rest area buildings. MDOT estimates a 30% reduction in annual energy costs for rest area lighting alone.

5.2.1 Administrative Functions (for Many Governmental Units and Many Functional Elements of Government This Is the Largest Contributor to the Primary "Product" of Service to Constituents: Business, Private, and Charitable)

5.2.1.1 Management of Delivery Models, Understanding of Best Business Practices, and Delivery Models and Modalities

The earliest stages of service delivery are analogous to the earliest stages of product development. They include product concept, initial product considerations, and, in a very loose sense, initial product delivery considerations. The customer organization interface including place of delivery, delivery facility "persona," and, trade-off between automation e-commerce services and live service would be the major issues. There are few if any life cycle management issues to be discussed as part of this, but decisions here can drive life cycle costs for hazardous and nonhazardous streams. In essence, government itself is held to a higher standard. The physical entity, as well as the social entity residents know as the "local government," must set examples in every facet of operation and in philosophy of operation. That goes beyond delivery of services at the lowest cost. The perception is not only of best value but sensitivity to all issues is part of the equation here, much more so than in any private enterprise.

5.2.1.2 Feedback From Frontline Staff and the Public

Closed systems work because the feedback leads to correction, thus increasing flexibility and encouraging resiliency. What one must remember is that an initial diagram shows a process with a beginning and an end; we close that loop by providing feedback, and that is what makes the system or the product better. Feedback is often used at intermediate steps, but the most telling, and in some senses the most important feedback, is feedback that is generated at the far end of the loop. Nowhere is that better demonstrated than government, when government works properly. For elected officials, that feedback comes regularly on election days. For us, the much more important issue is feedback from the customers, which is the public, and that is why frontline employees of government at all levels are perhaps the most critical element in the feedback loop. Government is unique in that the majority of the services are provided directly to end users by its own employees. Frontline employees of government are in a unique position to provide real-time feedback and correct many problems. The larger issue today is the inability of government bureaucracies to understand that it is the frontline employees who are the most critical component in the delivery of services to the public, and it is those frontline employees who routinely see the foibles of the government unit and the frustration of the government customers. Those customers are the public at large, whether representing themselves as private individuals, representing their business entities, or representing special interest groups.

Here is a real-world example of how things *are* and how things *could be*:

> A city planning board is working on a comprehensive rezoning effort. The planning board meets two evenings a month and has "solicited input from the public." The concept that holding two meetings a month in the evening and "allowing" members of the public to speak for about 3 minutes is somehow going to encourage the public to participate is inherently flawed. Such a process fails to recognize key issues, including the fact that many cannot make nighttime meetings; even fewer will even know about

> such meetings; and for those who wish to attend, there is no guarantee they will get the chance to speak on any given evening. A separate issue that presents a much greater challenge to government but also helps to explain why the public has such a poor view of government is that this scenario is not designed for free and open exchange of ideas. Such a scenario is a forum to allow government to work its will on the people, which is in clear contrast to the defined role of government to respond to the will of the people. That format is a very reasonable format and the correct approach for routine planning board meetings, but it is not appropriate for a major comprehensive rezoning effort that needs public participation to succeed and win acceptance. For those who come to provide feedback once every five years or once in a lifetime, the process is both daunting and scary. You must "sign in" and you must get up in front of everybody, stand at a podium, and speak into a microphone when all eyes are on you and you are "on camera"; this format does not allow for an exchange of ideas or open discussions. In this setting, the public is merely allowed to have very sterile and scripted "input" to the board, and that is quite different from participating in the process.

The author thrives in that environment and enjoys public speaking. For the majority of our population, speaking in a casual environment with friends is easy, but most are highly uncomfortable being in front of a crowd of strangers and having to do so in an unfamiliar and not overly warm venue. Many might not admit it, but they are intimidated by the process described above and go out of their way to avoid it. Government at all levels has become obsessed with the concept of what we are calling "transparency," but transparency does not demand this very structured and very clinical approach.

On the other hand, if such a board wants to engage the public, it is possible to hold open meetings and capture them by allowing small groups to meet in the neighborhoods in a much less formal environment where the only focus for the discussion is the rezoning effort. Those meetings could occur over a specific period of time, not indefinitely, so that people in every neighborhood would get a chance to participate, on a weeknight or on a weekend night or on a weekday or a weekend day, and to meet with some members of the board. Since all such meetings are public and records and minutes must be kept, transparency is preserved. The board would be able to engage people in an environment comfortable enough to encourage dialogue. This is a classic example of government having the best of intentions but failing to recognize the reality or think "outside of the box" to foster and encourage actual interchange rather than limited opportunities to speak.

Citizen awareness is critical to both hazardous materials mitigation efforts and recycling efforts; the largest number of hazardous materials incidents occurs in private residences and involve cleaning materials. For that reason, well-advertised and truly open forums where information can be shared rather than pushed down are required. A similar approach would be legitimate for local jurisdictions to get control of household hazardous waste. This is an area the majority of the nonurban jurisdictions in the United States simply cannot address because there is insufficient knowledge and awareness.

5.2.1.3 Office Supply Quantities and Materials

Within the larger construct of administrative functions, office supply quantities and materials are an area that very few companies examine closely in terms of risk and vulnerability in regards to hazardous materials and environmental impact. While everyone recognizes the inherent economic advantages of bulk purchasing office supplies, very few recognize just how many are actually hazardous in nature or regulated as hazardous wastes for commercial disposal. This is an area that is far more complex than we can treat here, but to give you a rough idea, this includes all of the following products: all forms of fluorescent light bulbs, any mercury light bulbs, all batteries, most toners for copying machines, "whiteout," "dry erase markers," many whiteboard and other electronic product cleaners, and some but certainly not all adhesives used in an office setting. While many in the Western world might not understand or recognize "carbon paper," the very name should warn everyone that there is an inherent liability issue here and, in fact, real carbon paper is a regulated item for that very reason and represents multiple hazards in storage, in transit, and for disposal. The easiest but certainly not a foolproof way to determine what risks in this area a product might represent is to simply inquire as to whether or not there is an MSDS. If an item has an MSDS, there has to be a reason. Granted, in some cases the hazard may be such that your school or business does not need to worry about it, but it is still your responsibility and ultimately the risk or liability does fall back on the business if an untoward occurrence results in a suit.

5.2.1.4 Lighting, Heating, and Cooling

There are number of separate areas here, and some of what we talked about will be repeated when we address production issues. Technology is also changing at a rapid rate, so material that may be correct at time of publication no longer reflects the actuality in as short a period as six months to a year.

The United States as a nation and as a market is shifting from incandescent lighting to other forms of lighting. A lot of that shift has already occurred in commercial and industrial spaces, so legislation-driven changes will most heavily impact consumers and private residences in the current near term. In the commercial marketplace, fluorescent lighting has been used for years. What many may be unaware of is that fluorescent light ballasts containing PCBs, and, starters from earlier units represented regulated waste streams, although their inherent hazards could be totally ignored as part of the initial sales or installation process. What all should be aware of is that a "cottage industry" came into existence some time ago once the inherent hazards associated with fluorescent tubes came recognized. In the 1950s and 1960s, it is unlikely anyone recognized the looming need to design and manufacture low-cost but effective packaging to return burned-out fluorescent lights. Today that is the norm; those lights need to retain their integrity until they arrive at a properly designated and licensed HAZMAT recycling facility. That means breakage must be avoided at all costs. Today's current family of CFLs (compact fluorescent lights), which are rapidly replacing incandescent light bulbs, involve the same hazards that fluorescent lights have always represented, that is, primarily a mercury vapor hazard. That means there are liabilities as well as costs attached to the use of all forms of fluorescent lighting. Assuming that one does not or cannot obtain incandescent light bulbs (ILBs), the alternatives include mercury and halogen as well as LEDs. At this point in time, remembering the caveat above, regardless of the acquisition cost it would appear that LEDs provide the greatest opportunity to reduce liability and risk and to eliminate hazardous material waste streams. LEDs are currently quite expensive, but they have outstanding lifetime characteristics, and to the best of the author's knowledge represent no form of hazardous waste stream when they reach the end of their service life. Mercury light bulbs clearly represent the same hazards as all fluorescent bulbs; please note the word "mercury" in their name. Halogen lights have enjoyed and continue to enjoy a fair bit of popularity. From the life cycle, operation cost, safety, and ecological standpoints, halogens represent the least attractive alternate light source. Halogens tend to run at very high temperatures, and that alone makes them less desirable because the high temperatures they generate represent a large waste of energy and a theoretically higher risk as an ignition source that could start structural fires.

Let us talk a little bit more about lighting without becoming engineers or scientists. According to one *National Geographic* article, most incandescent light sources are between 1.9% and 2.6% efficient, while fluorescent light sources are typically 9% to 11% efficient, 3 to 4 times more efficient than incandescent (Moll, 2012). Current claims are that LEDs are 15 times more efficient than incandescent lights. That makes LEDs very competitive since there are no hazardous materials in the product, they run cooler, are more efficient, and therefore present less risk and liability both in operation and for disposal. In reality, incandescent light bulbs are heaters that also provide light; that is especially true when talking about halogens. What is overlooked in most discussions are the true life cycle costs for the different forms of lighting. In colder climates, where in the past incandescent lighting was providing incremental heating, a switch from incandescent to fluorescent can actually increase greenhouse gases, increase total life cycle costs, and therefore have yet another detrimental effect on the environment and the economy. Assuming one could continue to use incandescent lights and the facility was in the appropriate climate, such as those with four seasons, above certain altitudes over a vast portion of the globe or below certain latitudes, incandescent lighting might provide the best set of trade-offs. If that is true, a lot of research and a lot of engineering has to go into the actual facility to maximize benefits and minimize costs. With sufficient opportunity to help meet a heating load, even seasonally, incandescent lights may make a much better choice than business has been led to believe, and unfortunately what government and environmentalists think they understand. All sorts of facilities closer to the equator must look at LEDs as the preferred alternative, based on science and not the propaganda put out by many organizations, including governments. Switching from ILBs to CFLBs may not always result in an environmentally friendly outcome, especially in cold climates, as the authors of a Canadian report conclude: "In summary, switching from ILBs to CFLBs may not always result in an environmentally friendly outcome, especially in cold climates" (Ivanco, Waher, & Warbody, 2009).

Heating is another extremely complex area; for the vast majority of businesses it can be placed lower on the list. Realistically, individual operations are going to have to make a number of different choices about heat.

The primary choices are either electrical heating or some form of fossil fueled system. Based on the technologies available in 2012, choices for many smaller businesses in their own facilities are extremely broad. For those in leased or rented spaces, utilities may or may not be included in such payments, and therefore choices may be more limited. For those who are planning on using leased or rented spaces, consideration of heating and cooling sources should be a factor in selecting a location and might even need to be looked at from an economic versus environmental cost trade-off standpoint. For those who are actually going to design or build their own new facilities, both geothermal and solar systems offer reasonable alternatives either as primary systems or as auxiliary/backup systems. The bottom line is that from an environmental and life cycle standpoint, natural or liquefied petroleum gas offers significant advantages over all other forms. Today, additional sources are being developed that might replace any or all of these depending on the size of the operation, but all need to be considered and the positives and negatives in terms of hazards to the air hazards to the earth and waste streams have to be examined. The larger issue is what sort of operations go on in the business and whether there is a way to recover large quantities of otherwise wasted heat to offset the need to generate additional heat in order to maintain a viable internal environment. As an example, and a historic reference point, in the days before digital electronics, it was not unusual for military communications facilities to be designed to recapture the heat from tube-operated equipment to maintain reasonable operating temperatures within their facilities. The Naval Communication Station in Adak, Alaska had a very aggressive plan to recover heat from a lot of the tube-operated terminal equipment to help maintain the temperature in their facility and reduce heating costs.

Various forms of geothermal heating and cooling are also available, but there are many variables, which include water tables, quality of water, overall installation, and maintenance costs of such a system and the recognition that by strict definition heated water becomes a regulated waste stream because heated water pumped back into the environment can have significant impact, for groundwater. The impact may or may not be minimal for water pumped back into bodies of water, but the impact on the entire water bodies' ecosystem can be significant. We will come back and look at some additional opportunities for capturing "heat" associated with any number of operations. When we start talking about heat recovery and air distribution, something as simple as effective ventilation allows the recovery of a tremendous amount of heat value in production and related operations that in turn can help reduce the "costs" of heating.

Cooling offers fewer opportunities today simply because of the laws of physics and the current state of technology. However, geothermal cooling and simple wet cooling properly deployed in appropriate climates offer opportunities that are seldom considered, but with the rising cost of fossil fuels, increased regulation, and the economic cost of regulatory compliance, these are areas that are also going to need further study. One big issue with cooling when one gets beyond a certain facility size is that many current techniques use chemicals that are hazardous, and they become a challenge to manage, represent increased liabilities, and in most cases create hazardous waste streams, ammonia is a leading coolant.

For very large facilities or for larger properties such as schools or school campuses, large office parks, prisons, ports, and similar operations, it is not unusual for the occupying entity to build its own power generation facility and for larger properties, water treatment and sewage treatment facilities. Based on current technology, new forms of energy generation including such technologies as pebble bed atomic reactors might make sense. Many power generation processes depend on the heating of water, and that represents an additional set of risks liabilities and waste streams. Factors often overlooked in the development and management of heavy industries/large commercial facilities are the costs, dangers, of waste streams associated with boiler water treatment chemicals, scrubbers, and other peripherals and support operations required to maintain such a facility.

5.2.1.5 Utilities and Space

Except for fairly large facilities, utilities are most often purchased, so there is little in the way of hazardous materials or waste streams especially those waste streams of airborne or waterborne emissions. That leaves the issue of space. This becomes another set of balance or tradeoff equations: should one build up or outsource; can production or other inherently dangerous processes be isolated from or integrated into the overall space? How does building design and construction impact "livability" and adequate physical working space? Can effective use of any number of different techniques reduce the carbon footprint, the physical footprint, or the environmental footprint of a facility? One area that will receive more attention is the use of pervious rather than impervious materials for parking spaces. Another will be use of natural light to supplement electrical lighting, another is proper siting on a lot or proper selection of a lot to allow for proper siting and the need to plant or take advantage of existing tree lines based upon heating, cooling, and natural weather patterns. In essence, what we are saying is that any way you can reduce the need or use of generated power is a way to improve bottom line while at the same time providing for more environmentally friendly operation. Since we are taxing on the actual impact on the environment is not an area that should be ignored although in many cases it may be an area that falls outside of the organization's ability to impact or change.

5.2.1.6 Cleaning Supplies and Disposal Methods

Most people today recognize that the range of cleaning products and, to a lesser degree, such articles as oily rags represent another area of risk and open environmental contamination. There are far too many products that carry far too many risks for us to cover in this rather high-level overview. The primary issue here is that a very large number of cleaning supplies represent hazards of one sort or another, and there is an even greater danger in mixing the immiscible or chemically reactive cleaning materials. Many facilities that employ rotating machinery and use petroleum based lubricants do not discard oily rags and other materials that absorb lubricants or solvents. There are a host of companies today that either manage or recycle such industrial effluvia. The important point is to recognize the inherent risk associated with the materials and then manage that risk. See some of our discussions to follow.

In summary, there are far more hazardous materials and possible hazardous waste streams in the typical office environment and in the janitorial arena than most businesses ever consider, and most do not give enough thought to this. When one starts looking at the incremental volumes and risks from all operations, from the simplest office operations to the most complex production operations to the most difficult process maintenance procedures, the total volume of materials that would to be examined is huge. Chemicals and substances of all sorts are subject to regulation and therefore must be considered or treated as "regulated" materials. All are subject to some form of control, and when all areas are recognized, both the number and the quantity of such regulated materials grow quite large and that does, in fact, represent a measurable and quantifiable risk to the organization.

5.2.2 Internal Industrial Functions

5.2.2.1 Management of Design; Understanding of Practices, and Materials to Be Avoided

Large portions of this section would not apply to the vast majority reading this, but even if your organization does nothing more than maintain a fleet of vehicles, then there is value in a lot of this material.

The earliest stages of a development cycle for a product may go by many different titles, but they do include product concept, initial product considerations, and, in a very loose sense, initial product design. At this early stage, it is important to recognize the disadvantages and risks associated with many items that are often taken for granted:

- Understanding the inherent risks of newer technology products
- Acknowledging changes in technology that are eliminating materials that in the past would have had to be considered hazardous wastes
- Having a basic awareness, and understanding of, materials engineering, that is, composite materials instead of metals

Aesthetic issues—which may be major issues for consumer-based products, versus the costs of reaching satisfactory aesthetic results—also need to be factored in early in the product development cycle. There is no right or wrong answer, and any specific choice must be governed by many considerations not just the hazardous materials and environmental considerations. The issue here is to understand the inherent advantages and disadvantages of materials, processes, and design standards, as well as manufacturing production standards in order to balance out pros and cons for each individual product.

5.2.2.2 Product Selection Criteria Going Beyond Pricing

Once the designers and engineers have begun to develop an actual physical product, even in the preprototype stage, a logistician needs to get involved. The logistician needs to understand concepts from many different fields, including materials engineering, cost-benefit analysis, reliability and, of course, corporate risk because of the inherent hazard unique to a selected input stream. There is considerable room for new types of collaboration between the design team and the procurement specialist. Procurement specialists are going to have to gain new skills, and design engineers are going to have to give up some old skills for the system to function most effectively and efficiently.

Product selection must balance out additional unique factors once form, fit, and function have been addressed. Up-front acquisition costs, life cycle costs in terms of maintenance repair liability, and hazardous risks, whether that is during operation, including usage by the end user down to the private individual at the retail or home-use level. This also includes considerations related to life cycle disposition costs, the concept and relatively new field in 2010 of reverse logistics. Up-front cost or purchase price may be identified as one of the largest drivers, but it is no longer the only driver in selecting component parts. In many cases, price may be more important than the areas mentioned above, but in other cases higher-priced items may prove to be more costly in terms of product life operating and maintenance costs as well as product disposal costs. Society today, through international law and trade agreements, is beginning to levy a charge on the manufacturing or producing organization or present other perceived risks to the planet.

A brief discussion of three commonly applied plating processes demonstrates why this is truly a critical area today. Look at the opening news item at the very beginning of Chapter 1. Three standard industrial plating techniques are summarized below. Depending on perceived application and the technical needs, it is in fact the logisticians, as well as the risk management or environmental health and safety personnel who need to understand the direct and indirect societal environmental costs of choosing component pieces, parts, and fasteners that have been subject to the various plating techniques. This is an issue that needs to be addressed between corporate management, finance, engineering, sales and marketing, and logistics. This is not a one-size-fits-all or a single department decision-making process.

Cadmium plating (or "cad plating") offers a long list of technical advantages such as excellent corrosion resistance even at relatively low thickness and in salt atmospheres, softness and malleability, freedom from sticky or bulky corrosion products, galvanic compatibility with aluminum, and freedom from stick slip. Stick slip is important because it allows reliable torquing of plated threads. Cadmium plating can be dyed to many colors, has good lubricity and solderability, and works well either as a final finish or as a paint base. If environmental concerns matter (and they do today), in most cases cadmium plating can be directly replaced with gold plating as it shares most of the material properties. The major drawback to gold plating is that it is more expensive and cannot serve as a paint base.

Zinc coatings prevent oxidation of the protected metal by forming a barrier and by acting as a sacrificial anode if this barrier is damaged. Zinc oxide is a fine white dust that does not cause a breakdown of the substrate's surface integrity as it is formed. Zinc oxide, if undisturbed, can act as a barrier to further oxidation, in a way similar to the protection afforded to aluminum and stainless steels by their oxide layers. The majority of hardware parts are zinc plated, rather than cadmium plated.

Chrome plating. Typically, when we are using this term we are referring to what is colloquially referred to as decorative plating; think of chrome parts on a show car or bathroom and kitchen fixtures. This form is typically a 10-μm layer over an underlying nickel plate. When plating on iron or steel, an underlying plating of copper allows the nickel to adhere. The pores in the nickel and chromium layers also promote corrosion resistance. Bright chrome imparts a mirror-like finish to items such as metal furniture frames and automotive trim. Thicker deposits, up to 1000 μm, are called *hard chrome* and are used in industrial equipment to reduce friction and wear.

We give here a brief review of the growing environmental concerns about cadmium. At one point in time, asbestos was used universally as an insulator, which included such everyday items such as potholders and flooring tiles, to name just two. When DDT as an insecticide was first introduced, it changed farming in such a way that a much greater portion of the world's population could be fed from the same acreage of crops. For a number of years, carbon tetrachloride (CCl_4) was used as a fire suppressant and a solvent to clean electronic contacts, among other things. Mercury was used for thermometers, many electrical applications, and as a primary material in the formation of dental fillings (many of us remember our dentists turning pennies into "dimes" using mercury). Halogens and other chlorofluorocarbon (CFC) materials led to the introduction of automobile air conditioning and proved extremely valuable in fire suppression systems for spaces packed with computers and other electronic equipment. Halogen was a major step forward for fire suppression for electronic spaces because all the other then current suppression techniques made the very equipment they were designed to protect unserviceable and/or led to significant degradation of the electronics. Two growing controversies in 2012 are use of pervious surfaces, primarily for parking areas, and lawns. Yes, that is right, lawns; there is a growing body of evidence that in America at least, people's love for green lawns is significantly reducing water quality in freshwater bodies and has destroyed large portions of once rich shellfish and other marine life, rivers, lakes, and streams. The Hudson River, which for good portion of this length is the geographic dividing line between New York and New Jersey, was once known for its large oyster beds. In New Hampshire, the Great Bay and the Oyster River were also once homes to large shellfish beds, hence the name Oyster River. Shellfish were essentially wiped out in both these bodies of water a long time ago, and significant efforts are being made today to clean up the water to the point that commercially viable shellfish beds can be brought back. You will see the same issue along the East Coast of the United States extending beyond shellfish, although some of the causes are quite different from the nitrogen overload created by fertilization to give us green lawns. Today we recognize every one of these as presenting significant hazards to our individual health and the health of, or balance in, the environment, and many of these materials, or chemicals to be more accurate, are either no longer available commercially or are subject to strict and very limited use applications.

Another issue is that technology has demanded the use of more precious metals and other rare elements. Many of those are mined and recovered from a shrinking number of geographic locations in fewer and fewer countries; there is growing concern that this is both an economic and defense weakness in the global economy as it is allowing nation states to attain monopolies on critically needed materials. For those interested in reading more about this subject, CRS report R41744, "Rare Earth Elements in National Defense: Background, Oversight Issues, and Options for Congress," dated April 11, 2012, is an excellent analysis.

Hopefully, all this drives home the point that understanding the industrial processes behind components as small as an individual screw and as large and complex as complete electronic/computer printed circuit assemblies, which go into a product, is critical to multiple decisions. Those decisions include what parts to specify, what parts to make in-house versus what parts to outsource, and how design development and delivery of products impacts the environment and reflects corporate culture, mission statement, and goals. While at the same time, decisions about where to build, how to build, and what to build also have a direct impact not only on the environment but on the risk and liability of the "owning," or in many cases the leasing, commercial facility, whether that be a manufacturer or a law firm.

5.2.2.3 Chemical Processes, Chemical Selection, and Chemical Recovery and Treatment

There are two primary topics for discussion here. First is the choice of completing processes in-house or outsourcing; the second is the choice in the selection of products and processes to be used for component parts and possibly assembly and internal application of some of the most common processes. Most business majors and managers owners need to understand industrial processes such as printed circuit manufacturing and various plating processes and options, or spray painting versus dipping or other process given today's regulatory environment and social shift. Any takeover, expansion, or new business must reflect awareness on the part of the management team of the inherent risks and advantages of such types of industrial processes. The same holds true for logistics managers in existing operations; they must understand the inherent risks, vulnerabilities, and life cycle impacts of such industrial processes.

Printed circuit assemblies, whether they are called board strips or films, essentially depend on an acid etching process that removes copper from a coated surface. Medically, copper is recognized as one of the many heavy metals. It is critical to the human body, but it is also toxic to the human body and therefore a lessened copper waste in any form, including dissolved in acidic liquids, the better the environment. Unfortunately, the reality is that copper is the most effective and efficacious conductor of electricity, so we will continue to use, one for another the foreseeable future. That is not necessarily a bad thing, but overuse and unnecessary use of copper in terms of industrial processes and manufactured products should be a consideration. While there are no worldwide shortages of copper at this point in time, it is not logical, rational, or safe to assume that will hold true in another 20, 30, or 50 years. If copper becomes less available, the cost of copper will go up.

The acid etching process is the second part of the equation. By its very nature, acid etching consumes large quantities of acid and generates large quantities of waste liquids. The costs associated with the use of hazardous materials are far greater than most recognize, and represents an overhead or hidden cost not easily allocated to individual units of production. The acetic waste must be neutralized or treated whether that be on-site off-site. Realistically, the area of waste materials treatment has become too expensive; except for the largest operations, acid etching procedures are going to place fairly significant hazardous waste and disposal costs on third parties. Environmental health and safety costs that involve compliance training and facility design are considerably greater than most people realize. An excellent example would be a facility that was going to engage in a new, first-time acid etching process. Today that would require a considerable amount of personal protective equipment; very specific ventilation and circulation requirements including, by definition, acid-resistant materials as distinct from the traditional galvanized sheet stock that is used for the bulk of circulation systems in commercial and industrial facilities; storage separation and segregation issues; as well as effective fire protection and fire-fighting systems in areas where the assets are used or stored; additional training and training compliance costs; and, an often-overlooked aspect, facility design, which ensures leaks either in storage areas pre- and postproduction or in the production operations, are captured underneath the work spaces and not allowed to run into the ground or run into the normal sewage recovery or storm water systems. As you can see, there are considerable costs, risks, and liabilities associated with the production of printed circuit assemblies in the commercial environment.

Once the assemblies are created, components need to be mounted onto them and soldered. The choice of soldering material and the soldering technique have environmental, economic, and hazardous material life cycle dimensions. Spot soldering versus wave soldering is one consideration, while leaded versus unleaded solder is a second consideration. Do not assume one has an inherent advantage over the other.

The use of plated products is another little understood area. Very few operations today do their own plating, there are two simple reasons: plating processes today still depend on a chemical process that uses arsenic compounds extensively. This is simply too toxic a material for anybody but the "experts" to be using on a routine basis; that leads to the second reason for the cost for adequate compliance, the moment one involved in the use of such toxic materials comes extremely high. The larger decision beyond where plating should occur is whether there is adequate justification for plating as a technique today. This also tends to be an energy-intensive process that generates very large quantities of waste, and that waste often contains precious metals which need to be recovered. The actual plating location now has the daunting task of developing adequate and cost-effective separation tanks to recover materials and then handle what would still be partially contaminated highly hazardous waste materials. For lesser reasons, the use of plated products does need to be minimized across the commercial spectrum. A look at the trends in the commodities market, especially at the precious metals and rare earths segments so critical to modern solid state and high technology production, will shows that from a national security, and, for commercial production processes and many high-value high technology products the number of precious metals, and to a lesser degree other critical raw materials, are no longer available from the large numbers of sources they were available from less than 50 years ago. The sources for many of these critical building blocks for high-technology products is now concentrated in a very small number of nation states, creating a dangerous and possibly very expensive situation. Electrolyses plating is also available as well and it reduces the use of toxic chemicals, such as arsenic. It is a much slower process and cannot produce the same thicknesses. Such processes are more typically used in the decorative arts rather than in high-technology products.

As pointed out earlier, there are many different processes that depend on chemical action and reaction that include abrasion polishing and finishing processes to include painting or removing surface coatings from products, or simply preparing services further applications or treatments. There are also issues related to heat and electric conductivity and friction reduction. Proper treatment of surfaces made of the correct materials in with can reduce or eliminate the need for lubrication.

5.2.2.4 Material Usage and Selection

The previous two topics highlighted many of the issues related to material usage and selection. The two most important takeaways are (a) that one must understand the processes that create component parts of assemblies and/or finished products, and (b) EHS considerations, whether from the external production of purchased parts and components or from internal production processes, have significant financial implications as well as social responsibility marketing and risk dimensions.

5.2.2.5 Transportation Packaging

We will use slightly nontraditional terminology here in the hopes of maintaining clarity for all those who are not packaging professionals, and are not interested in becoming packaging experts. Technically, everything that contains, restrains, protects, displays, promotes, and adds to product usability or value at the user purchase level is "packaging." In a more focused framework, packaging is most often described as serving four functions and falling into three groups. The three groups generally used to categorize the entire packaging universe are: primary packaging, secondary packaging, and tertiary packaging. We are going to reduce that to two broad areas: primary packaging, that which occurs as part of the production process, and in many cases as an integral part of production (think of individual packets of salt, ketchup, or mustard, and of the individual dosage packaging of both over-the-counter and prescription drugs as examples where packaging is essentially an integral part of the product), and, transportation packaging, which is applied or conversely removed at the shipping and receiving areas of organizations.

Primary packaging may be an integral part of the production process or may occur at the very end of the production process so that it is an integral function of the product characteristic; and quite frequently initial packaging is literally contiguous to the end of the production process, so it will be addressed in the final segment (soda, beer, soup).

Logistics is involved with transportation and distribution, so transportation packaging will be addressed here. As part of the overall product development cycle and design process, there are issues that need to be addressed with logisticians and packaging specialists throughout the entire product development process. These decisions will directly impact on choices about specific technical functions for production line configuration, product package, product packaging, and transportation packaging.

> Transportation packaging has one overarching mission: preservation of the product from point of manufacture to the customer. This packaging adds to the value of the product in many direct and indirect ways. This includes the shipping container, interior protective packaging, and any unitizing materials for shipping. It does not include packaging for consumer products such as the primary packaging of food, beverages, pharmaceuticals, and cosmetics. In the United States, transportation packaging represents about one-third of the total purchases of packaging; the balance is attributable to primary packaging for consumer products. To design a transport package one must have goals or objectives in mind. These will vary with products, customers, distribution systems, manufacturing facilities, etc., but most transportation packaging should address the following:
>
> **Ease of Handling and Storage**—All parts of the distribution system should be able to economically move and store the packaged product.
>
> **Shipping Effectiveness**—Packaging and unitizing should enable the full utilization of carrier vehicles and must meet carrier rules and regulations.
>
> **Manufacturing Efficiency**—The packing and unitizing of goods should employ labor and facilities effectively.
>
> **Ease of Identification**—Package contents and routing should be easy to see, along with any special handling requirements.

Customer Needs—The package must provide ease of opening, dispensing, and disposal, as well as meet any special handling or storage requirements the customer may have.

Environmental Responsibility—In addition to meeting regulatory requirements, the design of packaging and unitizing should minimize solid waste by any of the following: reduction–return–reuse-recycle.

Since transport packaging should always be economical, the above goals should be balanced or optimized to achieve the lowest overall cost (McKinlay, 1999).

In 1914, American railroads, which at the time were carrying most of the freight in the United States, recognized and authorized the use of corrugated and solid fiberboard shipping containers for packing many different types of products. Motor carriers, in turn, followed the railroads' example in 1935 when they adopted their own packaging rules that often called for fiberboard boxes This standard packaging, employed during the early months of World War II, and specifically the colossal failures of packaging and the impact that had on the amphibious landings at Guadalcanal in 1942, which demonstrated how critical proper packaging design really is, forced the military to conduct their own research and create their own organizations to address packaging and eventually led to military specifications (MILSPEC) and military standard (MILSTD) packagings. The universally accepted packaging available at that time had proven adequate for rail and highway movement within the 48 continental United States but was grossly inadequate for the movement of goods overseas and into different and changing environments and prolonged exposure to the elements (Maloney, 2003). The scope of this disaster and the lessons learned led to the establishment of the School of Military Packaging Technology, the school disappeared from Aberdeen Proving Ground in October of 2009 as a result of BRAC, and some of this mission was transferred to the Defense Ammunition Center. Part of the reason this occurred was the rapid change in technology, specifically materials technology and packaging technology, which raised the overall standards of commercial packaging as well as military packaging.

The three biggest threats to product and packaging from a transportation and distribution standpoint in descending order are moisture, temperature, and pressure. If retail unit packages, which may contain multiple quantities of individual products or multiple packages of multiple quantities of products, include sufficient protection against moisture, there is less need for either more expensive for additional transportation packaging. On the other hand, as an example, if the organizational concept requires all outbound shipments be properly shrink-wrapped with the appropriate shrink wrap material, then theoretically the need for vapor barriers and desiccant materials may be reduced or eliminated. That might be particularly true for products that are produced for local or regional consumption and are yet still produced in relatively large quantities. This is why logisticians need a broad background than most recognize, and why logisticians need to be involved very early in the product development stages. Most texts on packaging identify four primary functions, although they may use slightly different terms or variance. The four functions of packaging are contain; protect; facilitate handling; and promote sales (Ramsland, 2000).

One area worth discussing briefly is packaging materials and packaging quantities. For those who entered the workforce in the 1960s and 1970s and have been involved with business-to-business commercial and industrial products, it comes as no surprise that packaging is a major issue. For those who study retail markets, consumer behavior, demographics, and psychographics, it is not news that packaging at the individual unit level for retail products presents serious challenges to getting the right balance between product protection and preservation versus safety versus ease of opening. The need to balance all three issues presents a myriad of challenges and has led to well-documented consumer frustration and dissatisfaction while adding to product costs as well as waste stream growth.

The many issues related to movement of goods from point of production to point of sale is another area that represents general waste issues, damage and loss issues, and environmental issues and impact. While not really addressed here, the whole new field of reverse logistics applies the moment goods leave the point of production. The more important point here is that the concept of reverse logistics must be engineered into the product, the production line, and the packaging function. This is necessary to address the growing legal ramifications associated with the production processes that generate hazardous waste streams and emissions as well as the ultimate ownership and responsibility of goods that are, or contain, hazardous materials.

The types of material used for what is known as secondary and tertiary packaging, the methods used to secure or "build" the packaging, and, the methods, processes, and materials used to generate visual information on the packaging all can add hidden costs that represent hazardous material exposures and increased packaging disposal costs. Some examples of specifics are adhesives used in building the physical package, or methods of closure/sealing; inks, paints, or other materials applied to the packaging to provide visual appeal or critical logistics and transportation information; preservation of the package itself application of fumigants to tertiary packaging and containers; and passive and interactive RFID tags.

One often-overlooked area that the text will not address in detail but is worthy of separate discussion is invasive species and environmental issues related to the movement of goods by sea from one ecosystem/environment to another ecosystem/environment that is geographically removed by great distances.

Kudzu, Asian carp, Zebra mussels, and long-horned beetles are just a few examples of invasive species that have drastically changed or are currently changing the local ecosystem. The most egregious example of such an unplanned consequence from modern air and sea transport is Guam; unfortunately it is one that very few Americans, including businesspeople, educators, legislators, and logisticians, are aware of.

It is now generally accepted that Americans unknowingly and inadvertently introduced brown snakes onto Guam in their post–World War II efforts to rebuild the island's economy. Those brown snakes have wiped out all native birds on the island, and it is now a full-time and not inexpensive job to manage and reduce the snake population on the island. There are ongoing efforts to restore an avian population to the island, but it will be an expensive and extensive long-term proposition.

While less pressing, those issues apply to commercial and noncommercial movement of goods and the transport devices (including those used strictly for pleasure) between areas within the United States. At the current time, because of an invasive species in parts of Worcester County, Massachusetts, raw forestry products, including firewood for personal use, cannot be moved out of the county for any purpose such as camp firewood per trip to another area. If you take the time to look at signs posted at many public boat ramps throughout the United States, you will notice that there is both a request, and a legal requirement, to properly clean or purge boats when they come out of the water even for movement within the local areas within the United States. An overlooked partner in this area is USDA APHIS. Agricultural products may not freely move into and out of every state because of the organisms they might carry, whether that is some form of plant disease such as blight or some sort of living insect such as arachnids or nematodes. Arachnids and nematodes love to travel in/on untreated wood, so we now have international phytosanitary standards. In the past, treatment of wood products was (or in some cases still) uses what we now recognize as extremely toxic/poisonous materials as an alternative to other forms of treatment.

Recent cases demonstrate that under certain conditions the treatment of such packaging materials used in transportation, such as pallets, dunnage, and blocking and bracing can contaminate the product being transported.

As you can see, there is a lot more to product packaging, packaging for transportation, and transportation itself that needs to be considered and discussed. We will come back to that later, but now let us move on to packaging directly related to the product rather than its shipment.

5.2.2.6 Material Handling Equipment and Material Handling Systems

If you have material handling equipment, then you have hazardous materials issues as well as environmental health and safety issues. This is not a bad thing, but if you have not properly addressed all the requirements or you simply did not understand that there were so many requirements, this could be a weak area for many operations.

All such items use some form of motor, which is rotating machinery, so immediately there are personnel safety issues and lubricant issues. Internal combustion engines, regardless of the fossil fuel, create additional series of challenges based on the fuels themselves, air monitoring requirements, and circulation considerations. Electric-powered units have batteries. Depending upon what kind of batteries are being used, battery charging and servicing stations as well as other unique EHS requirements that might include acid absorbent spill kits and neutralization materials, eyewash stations, open flame and heat restrictions, and possibly special fire suppression and fire-fighting materials and protocols are needed. All represent additional costs, risks, and exposures. Both industrial and small lithium-ion batteries have been known to cause fires, so adequate research needs to be done both on the batteries themselves and charging systems that will be used, including the possibility for the need of fire suppression and current limiting devices. If you use lead acid batteries to start your material handling equipment, then you will need eyewash stations and servicing stations with adequate ventilation and personnel protective equipment. For an operation using rotating machinery, including typical MHE lubricating products as well as lead acid batteries, are part of the package. There are two different sets of spill response/control items and two separate sets of requirements for safety materials and personnel protective appointment.

The larger issue is recognizing what hazardous materials are inherently parts of the material handling equipment. One category of hazardous materials must be kept on hand to service and support the material handling equipment and the kind of waste streams you are generating. There is a second set or group of materials that represent recognized and hidden waste streams. This includes everything from coolant to battery acids drained or unused excess oils and, of course, oily or otherwise contaminated rags. Solvents used to clean the equipment itself as well as the tools in the work areas and various preservatives may need to be treated as hazardous materials. Many service and maintenance aids that might not be treated as hazardous materials initially will have to be treated as hazardous wastes once they have been opened/used when disposing of them. That includes even the smallest quantities in the bottom of cans, drums, tubes, swabs, or brushes that have any foreign materials adhering to, or absorbed in, them. Some of the materials used in operating and maintaining material handling equipment are essentially incompatible, which means they must be segregated and that special efforts must be made to properly segregate those waste streams as well.

5.2.2.7 Transportation Methods

Most purchasers of transportation services and the vast bulk of the transportation operators, since the majority are single-mode operators, usually look at the cost, time, and convenience factors

Table 5.2 Modal Comparison Environmental, Safety, and Health Factors

Mode	*Energy efficiency index*	*Speed*	*Average haul*	*Deaths per billion passengers (kms)*	*Date introduced*	*Vehicle life (years)*	*CO_2 emission (gms per ton per km)*
Air	1	400	1000	0.02	1958	22	540
Truck	15	55	265	2.4	1920	10	50
Rail	50	20 (200)	500	0.55	1830 (1970)	20	30
Barge	64	5.5	330	Very small	17th C	50	
Pipeline	75	4.5	300	Negligible	1856 (1970)	?	
Liner	100	16.5	1500	1.0	1870 (1970)	15	25

Source: Alderton, P., *Port Operations*, Informa, London, 2008.

when making decisions in regard to mode selection. Today, informed businesses do recognize all the impacts of life cycle management and are also looking at other costs and impacts in deciding on their total distribution chain even if that function is exclusively provided by 3PLs and outside contractors. Energy efficiency and CO_2 emissions are both factors that need be applied when designing the distribution system. A properly designed system can minimize costs to the company in relationship to inventory cost; convenience for the customer, or rapid delivery; and green concepts or impacts. The chart (Table 5.2) clearly demonstrates that there are a number of factors that need to be considered by both transportation sector providers and users when making decisions in regards to transportation and distribution models.

For many reading this, a lot of the information in the table above is superfluous to the point of overkill, but for certain industries and different specific businesses and sectors all of the information has value and impact.

Realistically, very few modes of transportation can provide for uninterrupted door-to-door movement of goods from place of production to place of sale or use (the two are not necessarily the same when dealing with retail products). That means almost all goods will complete some part of their journey by truck. That by no means suggests that truck transportation is the best, or even a good, choice. In addition, if one understands the transportation industry, even for a material that travels exclusively via truck, it is likely that the material will flow onto at least three and possibly as many as seven or more different transport units before arrival at its final selling point. Many producing organizations capture the major issues related to operating their own fleets or contracting out to 3PL or other external provider. What few recognize are the huge chemical costs related to the service and maintenance of a vehicle fleet. All information that applies to material handling equipment above applies to an organization that wishes to operate its own vehicles or decides to maintain such a fleet.

It takes a careful examination of the information in the table to make decisions. And for many, the decisions are going to be driven by other factors. It is only when we start aggregating large quantities and talk about initial distribution that a lot of this information comes into play. Nonetheless, it is critical that the issues represented by the information in the table be considered when making large-scale, and to a lesser degree, smaller-scale or local, transportation mode decisions.

Many travel booking services now provide CO_2 emission figures as part of the reservation package for air travel; this is shown as how many tons of CO_2 emissions each flight leg of a trip represents. Given the direction of global regulation and taxation, the concept represented by that information now carries an increasing direct economic burden, whether that is viewed as a hidden cost, and, whether or not this is passed on to the ultimate consumer of the good or service. There are inherent safety and cost considerations related to each mode; based on either, it is fairly easy to create ordered lists to help choose transportation modes for the movement of goods. Reliability is another parameter; however, two parameters are becoming more important: energy efficiency and emission generation. With careful planning and attention to detail, the supply chain manager can build in improved energy efficiency and reduced emissions to reduce the cost to the producer and consumer. At the same time, wise choices and careful planning within the supply chain also contribute to the reduction of environmental impact. In the end, based on current trends, society as a whole is going to have to understand and then be willing to "pay for" the environmental impact of the various modes of transportation, as well as the environmental impact of the various modes of production and packaging. Businesses are going to have to accept and understand the negative impacts of industrial processes, which include production and transportation, upon the environment and then decide how to either avoid or allocate those costs.

While there are no absolutes, some basic decision points can be presented here. For those concerned strictly with local operations, the chart above has little practical application. For those who are shipping strictly within a region, the chart begins to take on meaning but the options and choices are still limited. For those distributing nationally or globally, the chart begins to take on a lot more importance. Realistically, both those who are shipping and shipping companies themselves need to pay particular attention to the first three columns and the last column in the chart. In addition to that, companies distributing beyond the region must factor in the issues of warehousing and distribution centers. Today, the majority of business texts treat the transportation distribution model as a trade-off between transportation costs and distribution center/warehouse costs and convenience. The big issue is the cost of inventory sitting in warehouses and distribution centers as well as cost of inventory in transit. That is an excellent model and is still the foundation for such discussions. Today, however, driven by the growing body of scientific research and the increasing tendency of governments to regulate and impose "carbon taxes" domestically as well as globally, it behooves both the business community at large and the transportation community at large to look more carefully at some of the information provided in Table 5.2.

Today, the final decision must be driven by a combination of company culture, interest in, and legislation encouraging "green" approaches to business; consumer perceptions; and, cost benefit analysis that trades off time versus delivery cost versus convenience. Make no mistake, for any entity that is delivering large quantities outside of the region, all those issues must be taken into consideration.

5.2.2.8 Transportation Providers and Protocols

If one looks at inbound logistics for goods-producing organizations, time, cost, liability, and reliability all become drivers. Often, adjusting overall production schedules may allow for a move to less expensive and less environmentally unfriendly modes of transportation. That is a key concept in this rather short section. The just-in-time approach to manufacturing or production does not necessarily or automatically negate the use of less expensive modes of transportation. Typically, less expensive modes of transportation also represent those modes with the lowest carbon footprint and environmental impact. The trade-offs include time as an absolute, reliability, and seasonal impact on the mode of transportation. Something as simple as scheduling pickups or deliveries at the manufacturing facility for "off" hours can have significant direct and indirect environmental and economic impacts. Time of day pricing for surface transportation is now embedded in the Southern California port regions and can be expected to expand throughout the United States in all major port areas.

Commercial trucking operations and rail operations can have significant impact on congestion as well as "pollution." If we parse the normal 24-hour day into three "shifts," one could make the case for shipping and receiving operations being restricted or focused upon what is traditionally labeled as the "midwatch" by the military. That could be an arbitrary 8-hour band of time, for example, from 8 p.m. to 4 a.m. Taking the theoretical step further, shipping and receiving areas can be physically isolated from the rest of the facility; then there might be additional heating and cooling cost reductions associated with operating such facilities when the rest of the facility is closed down. This is not to suggest that any work has been done to quantify the advantages and disadvantages of such a concept. Humans are "wired," if you will, to operate during the day and sleep during the night, so safety in terms of driver ability/capability is one of the many factors that would have to be looked at under such a scenario. From a pure transportation cost basis and a pure environmental impact basis, there are a tremendous number of positives to exploring such an approach.

In addition to mode selection and time of day decisions, another major decision for many organizations, one that should help drive the outsourced transportation function decision, is selecting an appropriately equipped and operating carrier. Here the issue is less a hazardous material or waste material consideration than it is an environmental impact safety and reliability issue.

The last piece of the transportation issue is packaging. While not a major consideration, the total distribution chain for product will drive packaging design and packaging decisions. Proper choice of primary and secondary packaging, coupled with the intrinsic characteristics of the product, may allow for consumer products to move with equal reliability and safety via any mode of transportation. On the other hand, larger industrial products that might not ordinarily be considered "fragile" might need additional packaging for ocean voyages, especially since today a very large number of such items are "modularized" so they may fit into shipping containers.

5.2.3 Business and Industry Activity Services

There is a knowledge and responsibility component that underlies the discussions in this and the next section. The knowledge component has to do with an understanding of some of the basics of the earth sciences. That includes just a bit of geology and meteorology as well as an understanding of the concept of floodplains, prevailing winds, and other natural hazards within, or impacting upon, the jurisdiction. Obviously, at the national level there are huge departments and agencies tasked with that responsibility; at state, tribal, and local levels, that is not always the case, but that does not relieve the government entity of those responsibilities. Subterranean and geologic issues such as underground caves and underground streams influence what can be built and where. Certain geologic characteristics work against pipelines, so in such a jurisdiction one would work against distribution of natural gas via pipeline and focus on delivery of propane via truck. In talking to elected officials and many involved in first response and hazardous materials mitigation at the local level, it is interesting to note how few have any idea of national or regional pipelines in their own jurisdictions. US DOT's PHMSA has an excellent application to help city planners, regional planning commissions, transportation planners, emergency managers, and first responders understand the inherent hazardous materials exposures in their communities. The public version of this site may be accessed through the link https://www.npms.phmsa.dot.gov/PublicViewer.

Fault lines and earthquakes are other natural hazards that have been overlooked for vast portions of the United States in the last half of the 20th century, but recent events worldwide have forced local governments to reevaluate fault line impacts on zoning and earthquake likelihoods on local planning and code enforcement. While not an issue in large portions of the United States, volcanic activity seems to have picked up in the last two decades with varying impacts and effects, so that too must be examined. Regardless of the existing codes, ordinances, zoning regulations, and code enforcement activities, the geography and geology of a jurisdiction help determine not only what can occur in the jurisdiction but where within the jurisdiction activities can be allowed to occur. Well water contamination is a major issue; therefore, an understanding of the actual aquifer system in the region is a bedrock responsibility of the local government entity and many of its component organizations.

Government at all levels has recorded, collected, and created data that it has organized and cataloged, thus making it available for its own functional requirements and for the use and benefit of its constituents. Those charged with direct or indirect responsibility affecting life cycle management of hazardous materials *need* to know what is available and where it is available. It is not necessary to know all this information, but it is critical to know where to find all this information.

There is a movement at the federal level to integrate key processes in emergency management into the area traditionally labeled as "planning," based on what we have learned in the last five years. The body of knowledge has expanded, new threats have emerged, weather patterns have shifted, and geologic events have increased in number and frequency as well as disposition over greater portions of the planet. Most cities and towns are now required to have an all hazards mitigation plan, and that too is a reflection the growing awareness of the breadth and depth of the "hazards" that jurisdictions face. For those reasons, mayors, selectmen, city counselors, city managers, emergency managers, planning departments, code enforcement, and zoning activities need a firmer grasp of hazardous materials life cycle management as well as key elements of emergency management. That level and scope of knowledge must now include ICS and NIMS. Just recently a new national disaster recovery framework has been announced. Many smaller disasters are, in fact, caused by hazardous materials, or perhaps more accurately, from the fallout from hazardous materials incidents; it is important for local, state, tribal, and national governments to understand not only hazardous materials life cycle but also emergency management and emergency response covering mitigation preparation preparedness elements. All of that has been factored into the discussions below.

5.2.3.1 Understanding Best Practices and Materials to Be Minimized or Eliminated in the Business Community

Before EPA and NIOSH were created, the universal approach to liquid waste streams was "dilution is the solution." For years, larger cities literally buried their solid waste or took it out to sea and dumped it, a variation on the dilution theme. Until fairly recently, we also accepted wood burning as a relatively benign process. If you go down to the plains of Canterbury in New Zealand, you can literally see the impact of burning wood for residential heat on localized environmental conditions over the plains in the wintertime; it is apparent to the naked eye. Many materials that were considered breakthroughs at the time of their introduction, or materials embraced as solutions to problems that were in common use as late as the 1960s, are now banned entirely or highly restricted and heavily regulated. This means that local governments in their oversight roles, which includes permitting and all forms of code development, and code enforcement, must have access to current science, regulatory trends, and best practices. The first two come directly from government at the national level, so knowing where to look and which agencies are critical to the drafting of codes and ordinances is critical. That must precede or drive proper enforcement and application of grandfathering or fees. Realistically, there are a limited number of US federal agencies one must be able to access for information relative to hazardous materials and hazardous wastes; in most cases, it is not necessary to even contact the agencies because the relevant information and data is available through agency public portals. In addition, it is important to keep track of material generated by the Congressional Research Service (CRS), General Accounting Office (GAO), and the National Academy of Science; within the National Academy of Science, the reports published by a number of different cooperative research programs within the Transportation Research Board (TRB) are particularly valuable. Realistically, the key federal departments would include the following: Health and Human Services, Transportation, Agriculture, and Homeland Security. Specific organizations that do not fall within the listed agencies include EPA and OSHA. In Health and Human Services, key agencies would include the Centers for Disease Control (CDC), the Food and Drug Administration (FDA), and the National Institute of Occupational Safety and Health (NIOSH); for Transportation it would be the Pipeline and Hazardous Materials Safety Administration (PHMSA); and for Agriculture it would be primarily Animal and Plant Health Inspection Service. This is certainly not the entire spectrum of resources relating to hazardous materials, but it is the most important in regards to all things related to hazardous materials life cycle management as it applies to local governance.

Often-overlooked resources for government, especially local government, are the professional associations and trade associations; best practices both in production as well as safety and response are often introduced through these organizations' sites, and if a new processing or manufacturing facility were coming into an area, one of the best ways to learn about the generic hazards as well as the best practices would be to identify a trade organization or industry site that would represent that type of facility. Some simple examples include the American Water Works Association, the American Trucking Association, the American Association of Port Authorities, the Chlorine Institute, American Association of Railroads, the American Petroleum Institute, and, many others. Here you also see the importance of developing strong working relationships between chambers of commerce and local economic development organizations, whether they are in the nonprofit sector or are organized within the local community.

The last piece of this equation is a clear understanding of the three elements that contribute to the exposure for a community. What occurs near the borders of your jurisdiction, as well as a clear understanding of what moves through the jurisdiction, is just as important as knowing what is generated within the jurisdiction and what wastes are disposed of or allowed to disburse within the region. For those who have never participated in, or reviewed, a properly executed commodity flow study, it might come as a surprise to discover that the real hazardous material threats for some jurisdictions are based solely on what moves through the jurisdiction on the class I railroads and smaller rail lines, major navigable waterways, and roadways.

5.2.3.2 Understanding, and Control of, Chemicals Used Across the Entire Commercial Sector of the Community

This goes back to the information covered above; the various professional associations and trade and industry associations are still the best source to discover what chemicals and processes are likely to be used by a company. MSDS sheets for products consumed or produced provide additional information; a jurisdiction should have a centrally managed and properly cross-referenced master MSDS database. Under the EPA, there are certain reporting requirements for hazardous materials, but that requirement represents a very small portion of the actual hazardous materials consumed in the jurisdiction and focuses on the most dangerous materials and the largest quantities; it does not represent a true cross section of what materials exist in the jurisdiction and in what quantities. While not mandatory, an understanding of the NFPA 7O4M standard and its application to all commercial facilities is one example of how hazardous materials may be better identified and managed at a higher level. A very brief discussion of this widely recognized identification system is worthwhile at this point. This is the four-colored diamond you see in many places. The four colors, starting from the top of the diamond and working clockwise, are red, blue, white, and yellow. There is typically a number in each with 0 representing no hazard and 4 representing a high hazard; three of the colors have narrowly defined meanings: red—flammability; yellow—reactivity; white—special hazard; and blue—health. The white area uses abbreviations and symbols to indicate unique messages that might include "acid," "alkali," "corrosive" (all three are recognized abbreviations, and typically only one of the three would be used), "oxidizer" (which is slightly misleading to the uninitiated as it uses the abbreviation OXY, which to most would indicate oxygen, which is only one of many oxidizers); water-reactive; biohazards; radioactive; explosive; and poison or highly toxic.

As a government unit, it is critical to make sure all commercial uses of hazardous materials are captured so that appropriate plans can be created and appropriate response actions instituted in the event of an incident. Relatively small quantities of more commonly used hazardous materials can present unique risks when mixed with other chemicals, or when used as accelerants for other chemical reactions, including fire. Therefore, the broad challenge is to capture the presence of all hazardous materials in the workplace. That specifically includes a number of retail operations that often get overlooked by the uninitiated. Grocery stores, drugstores, cleaners, beauty salons, barbershops, do-it-yourself and hardware stores, as well as restaurants, all use hazardous material. Dangerous quantities exist in all these locations in the event of a fire or other incident. Ammonia is the most common commercial refrigerating agent, and it presents a number of unique risks and response actions. Knowing what is being used and where it is stored or used within specific business, or for larger quantities such as propane gasoline or diesel where the storage tanks are located and how they are connected, is critical to overall life cycle management and emergency response. Often it is the interaction of two or more chemicals after an incident has occurred that creates the problem. In some cases, it will also be necessary to decontaminate both people at the site and the first responders "working" the site.

The intersection of the Planning function, the Emergency Management function, and Life cycle management of hazardous materials is the master plan and the zoning framework/ordinances.

5.2.3.3 General and Specific Recycling Programs

What was simply called trash up until the latter part of the 20th century is now recognized as something much more complex. Science and technology have also taken us to a point far beyond where we were even two generations ago. Technology today depends heavily on a large range of processes that use hazardous materials and generate hazardous wastes. Much of what we have called trash in the past is now considered recyclable material, and many of the materials that are legitimately thrown out as trash, as distinct from all the materials we identify as recyclable, do, in fact, contain materials that we now understand represent hazardous waste streams. For businesses, that complex dynamic has been recognized for roughly 40 years or more, but all of the complexities are not necessarily recognized or addressed below the state level of government. Awareness and training are sorely lacking across a broad spectrum of disciplines within government and within business. Today, both hazardous and nonhazardous wastes are subject to or eligible for recycling. The requirements are established as a result of multiple laws and the ensuing regulations that have been promulgated. There are far too many to address individually, and today almost all jurisdictions are already applying some standards to the disposal of industrial waste streams. There are still unique opportunities to blend industrial and consumer waste streams because volume is an important component of cost-effective waste management. One core concept that applies to government, NGOs, and business is the concept of regulated hazardous wastes and their ownership. In simplest terms, under the current system in the United States, hazardous waste must be declared as such by a generator. That part is pretty simple and straightforward; what is complex is the ownership and the oversight of that waste. The key point is that whoever declares material hazardous waste, owns it in perpetuity. The courts have muddied this concept by repeatedly issuing contradictory rulings. The courts often and regularly hold current owners responsible for cleanup of contaminated sites, and in some cases hold government at multiple levels responsible if the property has been abandoned. At other times prior owners who can be identified have been held liable. Given the large amount of ground contamination that has already occurred, this will remain a problem for some time to come; it is important to recognize that cleanup costs may have to be borne by local jurisdictions. Moving forward, government can take a more proactive role by better understanding the nature the processes and industrial facilities in their jurisdictions and the nature of the wastes generated, identifying those that need to be overseen, and carefully monitoring those waste disposal operations. *Brownfield remediation* is a term many have heard but few understand. The main issue in the 21st century is preventing the creation of any new brownfields, thus eliminating the need for future or new brownfield remediation. This boils down to a sufficient level of knowledge on the part of all levels of government and an effective education and awareness effort for government and industry. There is a widely held belief that business as a whole places profits above Environmental Health and Safety issues and willfully breaks the law and that government is "in bed" with business and does not enforce the regulations. That is simply not true, although there will always be a small parentage of individuals on both sides who will knowingly act unethically. At this point in time, and probably until the middle of the 21st century, brownfield remediation will be an ongoing effort. The EPA brownfields site updated in February of 2012 makes the following claims: "The Brownfields Program creates many benefits for local communities: Projects leveraged $18.29 per EPA dollar expended; Leveraged 74,809 jobs nationwide; Can increase residential property values 2 to 3% when nearby brownfields are addressed; and, Promotes area-wide planning" (US EPA, 2012). This program demonstrates yet another valuable resource for government at all levels; there are dollars involved, and work effort involved, that the federal government is willing to underwrite or cost-share with local communities. If the elected officials of the communities do not understand this, then they are ill-equipped to take advantage of these opportunities.

While larger organizations have access to both the financial and knowledge resources to set up their own recycling programs, many small businesses, especially small local family businesses, are seldom aware of requirements and opportunities. As pointed out previously, all fluorescent lighting, all batteries, many office products such as toners and ink cartridges as well is whiteboard markers and correction fluid, are in fact regulated and must be treated as regulated wastes. Appropriate collection, through community recycling efforts, can actually provide income streams rather than appearing as costs, and therefore a much broader range of recycling activities across a much broader spectrum is required. Programs in place at many national retailers are either not known to local levels of government or not understood by local levels of government. There are opportunities to involve veteran service organizations, other community service organizations, after-school programs or school clubs, and organizations such as the Boy Scouts and Girl Scouts in permanent cyclical recycling and recovery efforts. If the jurisdiction is consuming energy, then it may be possible to examine ways to recover both hazardous and nonhazardous waste to be used as feed stocks for any number of purposes.

5.2.3.4 Education Programs and Incentives

Understanding the direct real costs of what Bierma and Waterstaat label the chemical beast to government at all levels as well as the indirect costs to the community at large is the first step in designing and incentive or disincentive programs. For all levels of government below the federal level, the resources available from the federal government are seldom understood or appreciated. Nonetheless, there are many programs and resources available at little or no cost to help educate the public and raise awareness in a number of areas related to the life cycle management of hazardous materials. In some cases, there are convenient tools available at web portals. In other cases, very complex programs and tools are available to dramatically raise both public and local government awareness and compliance; even when they require extensive training, the potential overall cost savings can be significant. There are many grant programs available in this area. Some that come to mind immediately across a number of departments, or in some cases on a regional basis, include the FEMA Region one STEP program; US DOT PHMSA outreach program; EPA training on how to handle, and EPA funding available for drug lab remediation and cleanup; US DOT hazardous material program grant money; brownfield remediation training and dollars; programs such as CAMEO and ALOHA HAZUS; and many more. The larger issue is first educating managers and elected officials so that they are aware that such resources even exist. In any jurisdiction that is using fossil fuels for static operations, there is a possibility of capturing all the oil from the hospitality industry and perhaps smaller machine shops and automotive repair shops in order to convert to a new mixed-use high-temperature furnace that will allow the use of all those waste streams to reduce the purchased fossil fuels. If that were to occur, here are some of the advantages to the community at large:

The cost of fuel for the jurisdiction will go down.
The cost for removal and processing of many waste streams will go down and that is spread across all commercial users.
It would be theoretically possible to look at a recycling program for household animal and vegetable fats and oils.
If high enough temperature furnaces are used, that would allow some of the industrial hazardous wastes consisting of entrained hazards within otherwise benign lubricating oils to be burned.
The number of "ton miles" of hazardous waste movements will be reduced drastically, and greenhouse gases from the municipal furnace will be reduced.

This is not a simple task if you examine all the parts that need to fit together and work together, but it is certainly doable in some jurisdictions.

Whether you wish to call them user fees or taxes, there is definitely a trend toward all levels of government to assess costs on the consumers of fossil fuels and producers of liquid, solid, and gaseous "waste" streams. Local government needs to look at the tools they can use to move commercial sector operations in their jurisdictions into less damaging processes and procedures. Since there is a need to generate revenues to cover the costs of governments, this is a fairly complex issue. Giving a real estate tax incentive to commercial entities for the use of gas over oil might be worth examining in light of the real economic, as well as the environmental, costs of the use of one fuel over the other. The real challenge is in understanding or quantifying all the costs involved in the use of various fuels versus the costs to the jurisdiction from meeting environmental requirements driven by the clean air act and many other sources. We are talking about offsetting or reducing air and water pollution; extra costs imposed by, or funds withheld by, the federal government; and treatment costs on the part of the jurisdiction or municipality. Currently, two major issues in many areas are greenhouse gas emission levels and their impact on air quality, and nitrogen levels and their impact on bodies of water and the underlying aquaculture including, for example, eel grass. Compliance with federal standards and state standards is adding millions if not billions to the costs of treatment plants at the local level. Part of the total approach to life cycle management of hazardous materials would be to examine how to reduce the generation of waste streams while understanding that we will not be able to eliminate them entirely. Incremental reductions across large and disparate areas within the community can help reach regulatory standards without necessarily involving additional costs for treatment equipment, whether that be filters; a user gas tax/fee; or newer, bigger, and more expensive municipal wastewater treatments plants. Because we now know that water and energy resources are not infinite, especially within a given geographic and political footprint, the traditional model that equated higher usage with lower per-unit costs may soon have to be reversed, and this is one of the first areas that needs to be examined at the local level. Adequate knowledge of current science and best practices, whether in municipal government or private enterprise, becomes a key element to effective regulation and management of hazardous materials through their life cycle. Today, science has proved that neither solid waste disposal nor dilution using large quantities of water are acceptable practices to manage wastes in the 21st century.

5.2.4 Private Citizen Activity Services

5.2.4.1 Understanding the Hazardous Materials Challenges of Materials Used in the Home

Again, the issue here is an understanding on the part of the jurisdiction of the hazardous materials dangers in the home, hazardous materials waste streams coming out of homes, the current federal and state regulations regarding these materials and the need to integrate emergency management and life cycle management of hazardous materials into all education programs, including formal and informal programs. The classic example is the mixture of two common household cleaning products: bleach and ammonia. When combined, toxic, often deadly, fumes are generated. Today, there are growing concerns about small lithium-ion batteries; that concern has changed practices for pediatric emergency care. Yet another area that has been mentioned repeatedly here is a need to make people understand that all forms of fluorescent light bulbs, especially the new CFLs that many think are legally mandated to replace incandescent bulbs, need special handling and cannot simply be thrown away in the trash. That underlying requirement holds true for all forms of batteries today and many other household products, including consumer electronic waste, all of which either cannot or should not be disposed of into municipal, or private, landfills, or worse yet, used as solid fill for the citizen or the local business.

5.2.4.2 Effective and Appropriate Regulations and Ordinances

The first piece of the governmental management issue is an internal but solid and broad-based understanding of the household hazards that exist and current effective and best practices that can be deployed. Effective regulation must recognize the need for efforts beyond prescribed or regulated fiats so that use of hazardous material in and outside the home, and therefore the need to manage hazardous materials, is minimized. As science helps us better understand the impact of nonpoint source pollution, there will be a growing need to examine the use of many outdoor garden chemicals as well as the need to examine private water sources and septic systems. Regulations and fees associated with what is euphemistically referred to as solid waste disposal, as well as accessibility to such sites and hours of operation, all fall under the area of regulation and ordinance.

5.2.4.3 *Effective and Accessible Recycling Facilities*

This goes far beyond a jurisdiction's municipally owned, or for that matter, a privately owned solid landfill site. Public–private partnerships and cutting-edge technology now allow much of what is considered hazardous waste, as well as the more benign solid wastes, to be turned into profit streams, but that requires knowledge and effort. Separation at curbside is part of this undertaking, but partnering with larger retailers and other appropriate segments to help aggregate hazardous wastes, without adding significant cost to the retailer or the consumer, will be an important part of this effort. On the surface, every retailer who sells light bulbs or batteries should have a program to collect those waste streams from all individual consumers. The logistics from a practical standpoint make that an impossible standard. More realistically, service organizations, individual governmental units such as law enforcement and fire, and private organizations such as the retailers need to work together to make it easy and essentially cost-free for consumers to recycle all wastes, but especially hazardous wastes.

5.2.4.4 *Effective and Meaningful Outreach and Education Programs*

At the current time, the lack of knowledge within governmental units primarily below the state level, rather than a lack of resources, is the biggest obstacle to effective life cycle management. Local government knowledge and programs, however, cannot be effective unless there is an increased focus placed on education and outreach. This requires public–private partnerships. Chambers of commerce, realty boards, service organizations of all kinds, and major businesses within the jurisdiction need to work together in concert so that public awareness is raised and proper disposal of all hazardous materials becomes normalized at the family unit level. The K–12 schools especially need to learn, teach, and practice life cycle management and efficacious recycling. That is a huge undertaking, but the fact remains that there are a tremendous number of assets available through the federal government, through various trade and professional associations, and through a little innovation on the part of government, the private sector, and the myriad of social, ethnic, and service organizations.

5.3 Risk and Vulnerability Assessment, Liability Issues Mitigation, Minimization, and Prevention Strategies

Table 5.3 and Figure 5.3 help underscore the point that hazardous material exists all around us and that in general *very* few people have any awareness of, still less an understanding of, the hazards that surround us at work, at home, and in between. Since the public at large, the business community in general, especially the retail segment, and government as a multilevel institution do

Table 5.3 General Use and Household Hazard Identification Matrix

Material	*Class*	*Example*
Calcium hypochlorite	Oxidizer	Algicide, swimming pools
Sodium hydroxide	Corrosive	Detergent and soap (TLV)
Sodium hypochlorite	Corrosive	Swimming pools, laundry
Nail polish, polish remover, perfume/cologne, most stains, and some paints	Flammable liquid	Self-explanatory
Liquid flavoring extracts	Flammable liquid	Vanilla extract, lemon extract
Herbicides, pesticides, fungicides	Toxic	All those gardening chemicals
Fire extinguishers and hair spray	Compressed gas	Self-explanatory
Correction fluid	Toxic	Office
Toner	Flammable	Office
Ink pen cart	Flammable	Office
Printing inks (older formulations)	Flammable	Local sign and printing companies
Brakes	Asbestos	Cars and trucks
Smoke detectors	Radioactive	Home and office
Fluorescent bulbs and starters	Toxic	Everywhere (school)
Older fluorescent ballast (PCBs)	Class (carcinogenic)	Self-explanatory
Batteries	Corrosive and explosive	Everywhere
Markers	Flammable	In this class room
PCs, CRTs, media, hard drives	Toxic	Chromium, beryllium phosphors
Oxygen generators (all types)	Oxidizer	Aircraft and personal medical use
Airbags	Explosive and corrosive	Cars and trucks
Lithium-ion batteries	Flammable solid	Consumer electronics, toys, office devices, aircraft black boxes, and much more

not know or understand the risks, it is impossible to address the risks, form any risk threat and vulnerability assessments, or take any corrective actions. Many everyday items are hazardous material or can easily combine with other materials to create hazardous materials. *Chlorine bleach* and *ammonia* combine to produce deadly chlorine gas. Hydrogen peroxide as a common use antiseptic in a very dilute solution: 100% hydrogen peroxide (CLASS 5.1 OXIDIZER and CORROSIVE) must be chemically stabilized before it may be moved.

In the post-9/11 era, at least within the United States, most of this is pretty straightforward and relatively easy. It does not need to be expensive, but it certainly is not free. Without going into great detail here, the author believes the HLS-CAM product is the most effective tool to use to address this. HLS-CAM stands for Home Land Security Comprehensive Assessment Model. This is one of many competing products, and all function in pretty much the same manner. All of the programs the author is familiar with are quality programs, and you cannot go wrong using any of the nationally accepted programs/sources. The place to start identifying programs and how to go about bringing a program to the community/jurisdiction is with the Department of Homeland Security, as all such courses must be DHS approved and therefore should appear in their listings. The HLS-CAM program distinguishes itself because of its genesis. It is built upon a highly selective process that was used to create a highly competent, highly motivated but numerically limited teaching staff. In turn, that staff delivers the program locally with the overarching goal of making communities safer. It does that by preplanning and unique tailoring at the delivery site at the time of delivery; at by enabling participants from different and often competing "silos" to identify what threats and vulnerabilities exist. This interactive framework fosters private–public partnerships

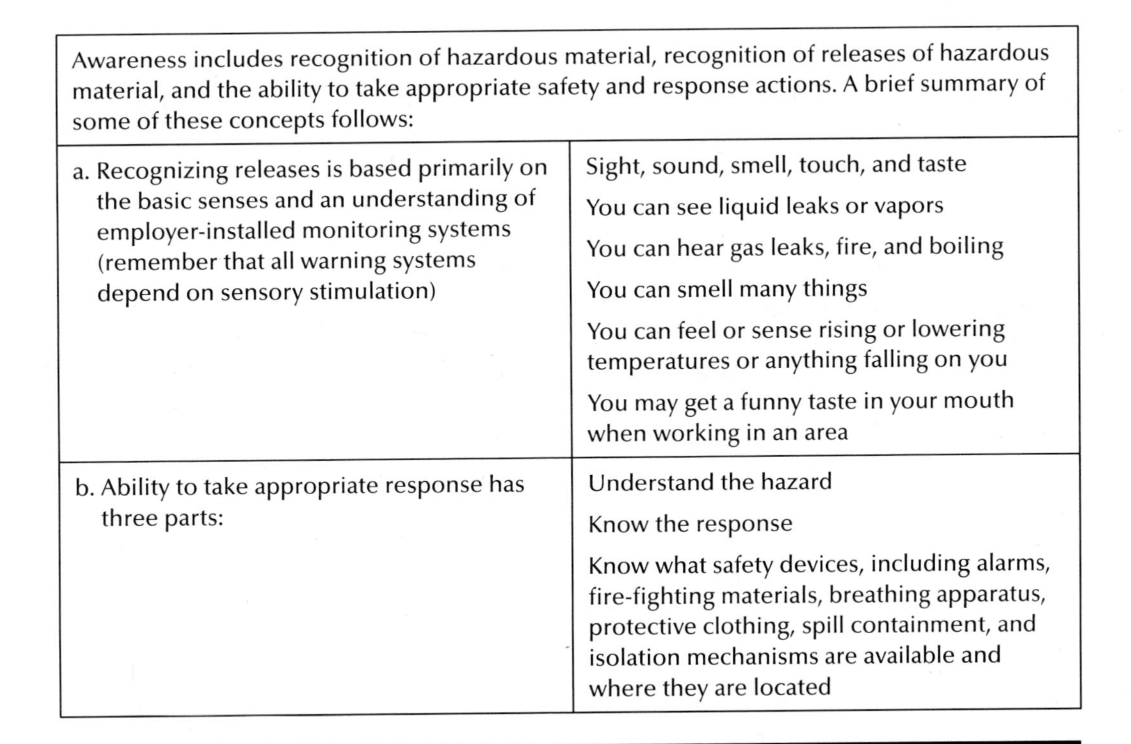

Awareness includes recognition of hazardous material, recognition of releases of hazardous material, and the ability to take appropriate safety and response actions. A brief summary of some of these concepts follows:	
a. Recognizing releases is based primarily on the basic senses and an understanding of employer-installed monitoring systems (remember that all warning systems depend on sensory stimulation)	Sight, sound, smell, touch, and taste You can see liquid leaks or vapors You can hear gas leaks, fire, and boiling You can smell many things You can feel or sense rising or lowering temperatures or anything falling on you You may get a funny taste in your mouth when working in an area
b. Ability to take appropriate response has three parts:	Understand the hazard Know the response Know what safety devices, including alarms, fire-fighting materials, breathing apparatus, protective clothing, spill containment, and isolation mechanisms are available and where they are located

Figure 5.3 Practical awareness.

and breaks down the silos or stovepipes within local government, initiating the first phase of open communication and long-lasting collaboration.

Liability in these areas is a large issue for government. In reality, the government sets the rules and enforces the rules on the one hand, while being subject to the rules and being responsible for communicating the rules to the citizens/residents on the other hand. There is no question that local governments regularly end up being held responsible for inaction as well as action, and often much of that could have been avoided had local government known and understood what hazardous materials existed and what remedies or oversights were available to protect the jurisdiction as a whole and protect governmental units specifically. In today's society, the perception is that in the end the government is somehow responsible for everything.

There are far too many variations from jurisdiction to jurisdiction to try and present any specific or even generalized mitigation minimization and prevention strategies. What might work in one climate will not work in another climate, what might work under one form of state government will not work in a different form of state government. It must be understood that there are very significant differences between state governments and the structure of units below the state level. The primary tools available are appropriate assessment; appropriate levels of knowledge within government; effective outreach and education programs across all citizens, including corporate citizens within the jurisdiction; and close attention to changing best practices and governmental regulations. In the end, government has the greatest liability of any entity; part of that is due simply to the fact that if real or personal property becomes abandoned, the "state" becomes responsible for that property. The other part of this misperception in most Western cultures is that the government through either action or inaction has responsibility as well as liability, and while that might not legally or always be true, it still comes back to the public's perception. For that reason, it is incumbent upon government to the lowest level practicable to understand all the issues related to hazardous materials life cycle management, and to engage in efficacious policies and processes that include education and outreach so that public awareness is raised to the point that government is no longer assumed to be totally liable and totally responsible.

5.4 Stakeholder Identification

It would be easy to say that everyone is a stakeholder when it comes to government, and there is a certain amount of truth to that idea. However, for the purposes of this discussion, there are a number of broad categories of stakeholders that all governmental entities must recognize.

Within the larger unit/level of government, there are its own employees as an autonomous group. That group in turn must be broken down into a number of subcategories, including the frontline employees who must interface with public on a daily basis as part of their job, those employees who have clearly identified impacts and lasting influence on the overall life cycle management of hazardous materials, and the senior management team within the entity, which would include elected, appointed, and paid individuals. Another closely related group of stakeholders would be external government entities at all levels who have influence on or are influenced by the unit itself in all those areas related to hazardous materials and hazardous wastes/emissions.

Outside of government itself, the major stakeholders are every household, and every and any business that either brings hazardous materials into the community, consumes and disposes of hazardous materials within or outside or the jurisdiction, or ships hazardous materials as finished products or as part of finished goods. That is still a very large set of stakeholders, which is one of the major challenges for government. In essence, when it comes to hazardous materials life cycle management, every person who lives or works in the community or travels through the community is a stakeholder. Those who live and work in contiguous jurisdictions are also stakeholders, and that presents a separate set of unique challenges.

References

Alderton, P. (2008). *Port operations.* London: Informa.

Bierma, T., & Waterstraat, F. (2000). *Chemical management.* New York: John Wiley & Sons.

Eartheasy. (2012). *LED light bulbs: Comparison charts.* Retrieved from http://eartheasy.com/live_led_bulbs_comparison.html.

Falkman, M. A. (2001, August 1). Packaging function, structure defined. *Packaging Digest.*

Griffin, R. D. (2009). *Principles of hazardous materials management.* Boca Raton, FL: Taylor & Francis.

Ivanco, M., Waher, K. R., & Wardrop, B.W. (2009). *Impact of conversion to compact fluorescent lighting, and other energy efficient devices, on greenhouse gas emissions.* Retrieved from http://www.intechopen.com/source/pdfs/11764/InTech-Impact_of_conversion_to_compact_fluorescent_lighting_and_other_energy_efficient_devices_on_greenhouse_gas_emissions.pdf.

Maloney, J. C. (2003). *The history and significance of military packaging.* Fort Belvoir, VA: Department of the Army.

McKinlay, A. (1995). *Need a new direction for your transport packaging? Do a 180!* Chicago, IL: IOPP.

McKinlay, A. (1998). *Transport packaging.* Chicago: CRC Press.

Moll, E. (2012.). *Energy-efficient bulbs: Halogen vs. fluorescent vs. incandescent.* Retrieved from http://greenliving.nationalgeographic.com/energy-efficient-bulbs-halogen-vs-fluorescent-vs-incandescent-3228.html.

Puleo, S. (2004). *Dark tide: The Great Boston Molasses Flood of 1919.* Boston, MA: Beacon Press.

Ramsland, T. (2000). *Packaging design: A practitioner's manual.* Geneva: International Trade Centre.

Threat Analysis Group. (2010). *Threat, vulnerability, risk—Commonly mixed up terms.* Retrieved from http://www.threatanalysis.com/blog/?p=43.

USEPA. (2012 February 6). *Brownfields and land revitalization.* Retrieved from http://epa.gov/brownfields/.

Chapter 6

Nonprofits and Nongovernmental Organizations (NGOs)

6.1 Introduction

In the United States, we use these two terms to describe organizations that provide routine services to our citizens on a daily basis as well as unbelievably large numbers of organizations that either provide services to affinity groups or exist primarily to provide postdisaster services. The more generally accepted term globally is NGO, and it is technically a more correct term because the connotative meaning of nonprofit is extremely misleading.

Because of the extremely wide variations in focus mission and public service, the following material will cover all possibilities, and therefore there will be major portions that are not applicable to specific organizations or organization-unique functions.

To maintain a common structure for this chapter, we have again picked the organization with four component parts, as shown in Figure 6.1. Just like the government model, the second depiction, Figure 6.2, is the more accurate representation of the operational flows for an NGO. The categories chosen for this chapter are

a. Donor operations
b. Administrative processes
c. Logistics operations and integration
d. Client operations

From both theoretical and practical standpoints, these are logical divisions, although, as in the government construct, in some organizations one or more of the functions may play minor roles for that particular organization. An organization providing outreach to the local community using MSWs would have relatively small donor operation and logistics operation components. Their primary focus more likely would be geared to getting contracts, identifying grants, running the administrative operation, and providing services through the local office or directly to their client base in the community. On the other hand, if one looks at the organizations such as the American Red Cross, all the elements become important, especially when one considers the major

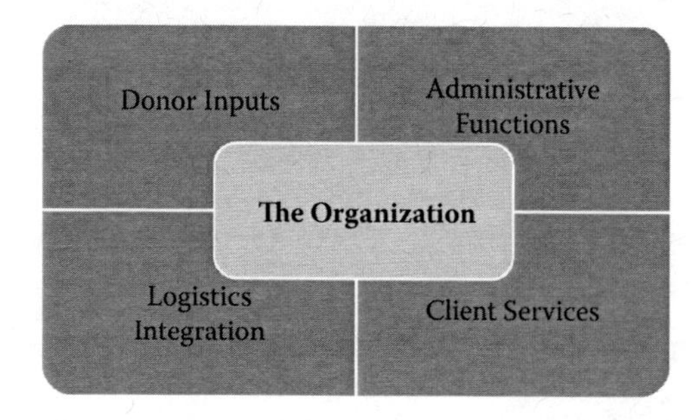

Figure 6.1 An NGO model.

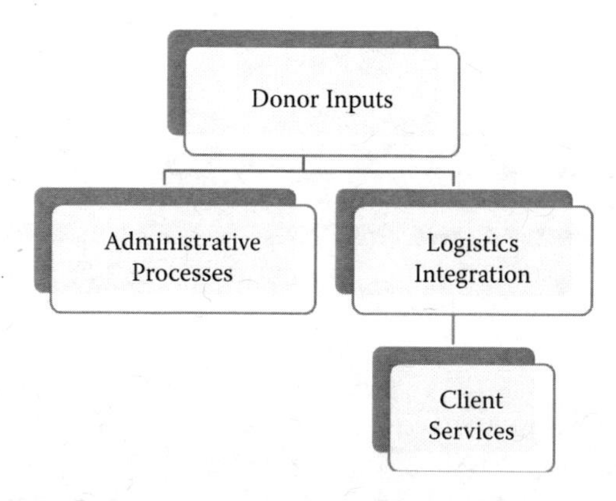

Figure 6.2 The NGO flow diagram.

role the American Red Cross plays in the American blood banking model. This is an example of an actual product development function as well as in primary product packaging function and an ongoing daily function that extends far beyond response to big or small disasters and emergencies. That places a larger or greater than ordinary burden on readers as they must understand their organization well enough to filter out the material that does not apply (Table 6.1).

6.2 Recognizing Defined and Undefined Hazardous Material and Hazardous Waste Streams

Figure 6.1 provides the framework for identifying common functions and the associated activities that generate the more obvious as well as the more subtle or harder-to-recognize waste streams. An aware and effective management needs to work at mitigating and/or minimizing all those streams and sources of emissions

6.2.1 Donor Operations

6.2.1.1 Understanding of Practices and Materials to Be Avoided

There are some very narrow issues here that would not apply to the other models in the previous two chapters. For many NGOs, collecting goods, new or used, is one of their primary functions. Such

Table 6.1 The NGO Construct

Donor Operations	**Exposure Risk Opportunity**
Solicitation of cash contributions Solicitation of goods and services Organization of volunteer efforts	Understanding of practices and materials to be avoided Understanding of unique issues related to collection of postpurchase goods that are, or contain materials that are, regulated as Dangerous Goods
Administrative Functions	**Exposure Risk Opportunity**
Administrative/office Facility management and maintenance Janitorial Support operations Purchasing of raw materials	Management of delivery models, understanding of best business practices and delivery models and modalities Feedback from public interface personnel Office supply quantities and materials* Lighting, heating, and cooling* Utilities and space Cleaning supplies and disposal methods*
Logistics Integration	**Exposure Risk Opportunity**
Product packaging Transportation packaging Distribution functions including storage warehousing transportation Product packaging Transportation packaging Distribution functions including storage warehousing transportation	Understanding of unique issues related to preparation and shipment of postpurchase goods that are, or contain materials that are, regulated as Dangerous Goods and the variations between nation states regarding such shipments and their posttransportation storage and distribution Transportation packaging* Product packaging Material handling equipment* Transportation methods* Transportation providers*
Client Services	**Exposure Risk Opportunity**
Normal service delivery Emergent need delivery of personnel, goods, and services	Understanding of regulatory environment and constricts in the field and nation state differences Training and credentialing of field staff and volunteers Understanding of practices and materials to be avoided in the field environment

* Material appears in Chapters 4 and 5.

collections are typically of postpurchase goods, an area that current literature would identify as within the reverse logistics spectrum; this is a year-round permanent activity. For others, collection of goods occurs fortuitously based on major disasters and emergent requirements; in some cases, those emergent requirements may recognize, or be in response to, chronic situations, but many NGOs attack these on a regular basis driven by trigger events. In all cases, the same process needs to occur: collecting usable materials, receiving or aggregating and processing those materials, and packaging and preparing those materials for shipment. Since we are dealing with tangible product coming primarily from end users, a number of separate issues must be addressed. It is extremely difficult to solicit the public for donations and then try to tell them what should not be donated and why certain items can cause sponsoring agency problems and cost them money. Organizations that collect tangible goods need to find as many ways as possible to minimize the actual packaging of those goods as part of their input streams, and must also look for effective ways to prefilter items that might require special processing, separation, and either disposal, or disposal of the hazardous components. This is no small task, but it is something that is not being adequately addressed by the organizations that are not involved on a permanent basis in international—or, for that matter, beyond local regional—distribution of goods. From an institutional standpoint, the organization needs to find a way to communicate the very large number of items that are not acceptable for donation because of the challenges they cause for transportation. An alternative, although labor intensive, is to accept everything and then depend on a large staff to sort and separate so that hazardous materials can either be redirected, isolated, rendered safe for normal shipment, or shipped separately. If an organization does have access to sufficient free but trained labor, then anything can be accepted. A corollary to this is exploring the entire field of reverse logistics and turning donated items that cannot be transported into cash streams to support a specific relief effort through aggressive recycling and refurbishing efforts or possibly even through preplanned bartering exchange formulas. This is an area that appears to hold great promise. Being able to accept "anything" from a public eager and willing to donate but having the internal wherewithal, which includes policies, procedures, and negotiated partnerships with those who can accept such materials and turn them into cash streams for the organization or turn them into in-kind services and goods for the organization, offers the NGO the opportunity to run a more responsive and effective operation. In the end, if there are ways to redirect all hazardous materials back into the local area, through whatever means, that is far superior to being put in the position of having to properly package and prepare these items for shipment and negotiating with carriers. In some cases, a carrier may not be legally able to accept dangerous goods for movement out of the region; this is especially for international movements even when using military aircraft and charters.

6.2.1.2 Understanding of Unique Issues Related to Collection of Postpurchase Goods That Are, or Contain, Materials Regulated as Dangerous Goods

Table 6.2 demonstrates just how many household products the general public takes for granted are, in fact, hazardous materials that are regulated for movement/transportation. The very large number of such items, when coupled to the lack of awareness of, still less an understanding or appreciation for, the hazards that surround us at work, at home, and in between is critical consideration for NGOs accepting postpurchase goods because there are many "hidden" hazards that a consumer does not have to consider or deal with. Unfortunately, every NGO, as a shipper, must be aware of and manage all those hidden hazards. Many everyday items are hazardous material or can easily combine with other materials to create situations that generate hazardous incidents or regulated materials. *Chlorine bleach* and *ammonia* combine to produce deadly chlorine gas. Hydrogen peroxide is a commonly used antiseptic in a very dilute solution; at those levels it is completely unregulated: 100% hydrogen peroxide (CLASS 5.1 OXIDIZER and CORROSIVE) must be chemically stabilized before it may be moved. Many items may be unregulated because of the way they are initially packaged or prepared, but when aggregated in a postpurchase environment, they represent legitimate and multiple hazards. Perhaps the best example is the way many small items are shipped with lithium-ion batteries installed. In most cases, there is a tab that must be pulled out to complete an electrical circuit with the battery.

Table 6.2 General Use and Household Hazard Identification Matrix

Material	*Class*	*Example*
Calcium hypochlorite	Oxidizer	Algicide, swimming pools
Sodium hydroxide	Corrosive	Detergent and soap (TLV)
Sodium hypochlorite	Corrosive	Swimming pools, laundry
Nail polish, polish remover, perfume/cologne, most stains, and some paints	Flammable liquids	Self-explanatory
Liquid flavoring extracts	Flammable liquids	Vanilla extract, lemon extract
Herbicides, pesticides, fungicides	Toxic	All those gardening chemicals
Cooking, heating, or lighting devices fueled	Flammable	Coleman® lanterns, Coleman stoves, immersion heaters
Charcoal	Flammable solid	
Untreated raw wood	Invasive species	Firewood, volunteer constructed wood pallets, unapproved blocking, bracing, and dunnage
Fire extinguishers and hair spray	Compressed gas	Self-explanatory
Correction fluid	Toxic	Office
Toner	Flammable	Office
Ink pen cart	Flammable	Office
Printing inks (older formulations)	Flammable	Local sign and printing companies
Brakes	Asbestos	Cars and trucks
Smoke detectors	Radioactive	Home and office
Fluorescent bulbs and starters	Toxic	Everywhere (school)
Older fluorescent ballast (PCBs)	Class (carcinogenic)	Self-explanatory
Batteries	Corrosive and explosive	Everywhere
Markers	Flammable	In the class room
PC CRTs media hard drives	Toxic	Chromium, beryllium phosphors
Oxygen generators (all types)	Oxidizer	Aircraft and personal medical use
Airbags	Explosive and corrosive	Cars and trucks
Lithium-ion batteries	Flammable solid	Consumer electronics, toys, office devices, aircraft black boxes, and much more

In a new item with the battery installed but the insulating material in place, there is no hazard. In a postpurchase environment, even a new item that has had an insulator removed represents a very real hazard. Yes, there are a lot of contributing conditions that can create a situation where this actually becomes a problem. If you remember all the videos that circulated a few years ago about exploding and burning laptops, you should recognize the inherent dangers to goods with lithium-ion batteries installed—it was those lithium-ion batteries that shorted out and caused the fires. This is probably the most extreme example because of the highly destructive nature of the lithium-ion battery short circuit. The same inherent threat exists for any goods that contain such lithium ion cells or many other types of batteries for that matter, in addition to any inherent hazard those batteries represent. Proper packaging and preparation of such goods and proper product design coupled with normal shipping procedures address the bulk of the problems for new goods. None of that holds true for a product that has been opened and checked to make sure it is working and is no longer really new although it may be "unused." In addition, there are separate issues related to the concept of "shelf life," and not all goods are clearly identified with a "use-by" date; in many cases, it is component parts such as gaskets and O-rings that are subject to shelf life limitation, and consumers often would not be aware of which internal component parts would fall into this category. For very large-scale operations, where tons of materials are being processed, all these issues have a direct impact on both the nature of and volume of material aggregating. It also creates challenges for identification, separation, and recovery options and requirements, and thus represents the final phase or step for the life cycle management of hazardous materials. Reducing the space, weight, volume, and nature of wastes has a direct, as well as an indirect, impact on the total cost equation for hazardous materials.

6.2.2 Administrative Processses

6.2.2.1 Management of Delivery Models, Understanding of Best Business Practices, and Delivery Models and Modalities

The earliest stages of service delivery are analogous to the earliest stages of product development, and they do include product concept, initial product considerations, and in a very loose sense, initial product delivery considerations. The customer organization interface including place of delivery, delivery facility "persona," and trade-off between automatic e-services and live service would be the major issues. There are few if any life cycle management issues to be discussed as part of this segment unless an organization is going to be tasked with routine, even if infrequent, collection of vast quantities of materials for processing to support a major disaster. For those NGOs that are either focused on delivery of goods and services in a postincident environment, or for those NGOs that have a component that addresses those sorts of situations, logistics and transportation functions are paramount. That still leaves the need for a core of permanent staff who have the necessary knowledge to oversee and administer programs that may ramp up rather quickly, and it also demands that organizations have the breadth and depth of knowledge to create policies and procedures that will ensure the free flow of goods from the public to the agency and on to the client. That may very well include the necessity to filter out and redirect certain product streams or component parts of consumer products while requiring funding and planning for purchase of replacement components or consumables for those items.

6.2.2.2 Feedback From Frontline Staff and the Public

Closed systems work when feedback works and leads to correction, thus increasing flexibility and encouraging resiliency. What one must remember is that an initial diagram shows a process with a beginning and an end; we close that loop by providing feedback and that is what makes the system or the product better. Feedback is often used at intermediate steps, but the most telling, and in some senses the most important feedback, is feedback that is generated at the far end of the loop. No place is that better demonstrated than in government and NGOs, when either works properly. For NGOs, that feedback comes regularly as donations come in. For us, the much more important issue is feedback from the customers, which is the public, and that is why frontline employees of NGOs are perhaps the most critical element in the feedback loop. Government and NGOs share a model where the majority of the services are provided directly to end users by their employees and organizational volunteers. Frontline employees of the NGO are in a unique position to provide real-time feedback and correct many problems. The larger issue today is the inability of large NGO bureaucracies to understand that it is the frontline employees who are the most critical component in the delivery of services to the public, and to learn from those employees, who routinely see the foibles of management and the frustration of the customers. Those customers are the public at large, whether representing themselves as private individuals or representing their business entities, or, representing special interests groups.

6.2.2.3 Office Supply Quantities and Materials

Within the larger construct of administrative functions, office-supply quantities and materials are an area that very few companies examine closely in terms of risk and vulnerability in regards to hazardous materials and environmental impact. While everyone recognizes the inherent economic advantages of bulk purchasing office supplies, very few recognize just how many are actually hazardous in nature or regulated as hazardous wastes for commercial disposal. This is an area that is far more complex than we can treat here, but to give you a rough idea, this includes all of the following products: all forms of fluorescent light bulbs, any mercury light bulbs, all batteries, most toners for copying machines, "whiteout," "dry erase markers," many whiteboard and other electronic product cleaners, and some but certainly not all adhesives used in an office setting. While many in the Western world might not understand or recognize "carbon paper," the very name should warn everyone that there is an inherent liability issue here and, in fact, real carbon paper is a regulated item for that very reason and represents multiple hazards in storage and for disposal. The easiest but certainly not a foolproof way to determine what risks in this area a product might represent is to simply inquire as to whether or not there is an MSDS. If an item has an MSDS sheet, there has to be a reason. Granted, in some cases the hazard may be such that your school or business does not need to worry about it, but it is still your responsibility and ultimately the risk of liability does fall back on the business if an untoward occurrence results in an incident.

6.2.2.4 Lighting, Heating, and Cooling

There are number of separate areas here, and some of what we talked about will be repeated when we address production issues. Technology is also changing at a rapid rate, so material that may be correct at time of publication they no longer reflect the actuality in as short a period as six months to a year.

The United States as a nation and as a market is shifting from incandescent lighting to other forms of lighting. A lot of that shift has already occurred in commercial and industrial spaces, so legislation-driven changes will most heavily impact consumers and private residences in the current near term. In the commercial marketplace, fluorescent lighting has been used for years. What many may be unaware of is that older fluorescent lights contained PCBs and starters from earlier units represent a regulated waste streams, although their inherent hazards could be totally ignored as part of the initial sales or installation process. What all should be aware of is that a "cottage industry" came into existence some time ago once the inherent hazards associated with fluorescent tubes came recognized. In the 1950s and 1960s, it is unlikely anyone recognized the looming need to design and manufacture low-cost but effective packaging to return burned-out fluorescent lights. Today that is the norm; those lights need to retain their integrity until they arrive at a properly designated and licensed HAZMAT recycling facility. That means breakage must be avoided at all costs. Today's current family of CFLs (compact fluorescent lights), which are rapidly replacing incandescent light bulbs, involve the same hazards that fluorescent lights have always represented, that is, primarily a mercury vapor hazard. That means there are liabilities as well as costs attached to the use of all forms of fluorescent lighting. Assuming that one does not or cannot obtain incandescent light bulbs (ILBs), the alternatives include mercury and halogen as well as LEDs. At this point in time, remembering the caveat above, regardless of the acquisition cost it would appear that LEDs provide the greatest opportunity to reduce liability and risk and eliminate hazardous material waste streams. LEDs are currently quite expensive, but they have outstanding lifetime characteristics, and to the best of the author's knowledge represent no form of hazardous waste stream when they reach the end of their service life. Mercury light bulbs clearly represent the same hazards as all fluorescent bulbs; please note the word "mercury" in their name. Halogen lights have enjoyed and continue to enjoy a fair bit of popularity. From the life cycle, operation cost, safety, and ecological standpoints, halogens represent the least attractive alternate light source. Halogens tend to run at very high temperatures, and that alone makes them less desirable because the high temperatures they generate represent a large waste of energy and a theoretically higher risk as an ignition source that could start structural fires.

Let us talk a little bit more about lighting without becoming engineers or scientists. According to one *National Geographic* article, most incandescent light sources are between 1.9% and 2.6% efficient, while fluorescent light sources are typically 9% to 11% efficient, 3 to 4 times more efficient than incandescent (Moll, 2012). Current claims are that LEDs are 15 times more efficient than incandescent lights. That makes LEDs very competitive since there are no hazardous materials in the product, they run cooler, are more efficient, and therefore present less risk and liability both in operation and for disposal. In reality, incandescent light bulbs are heaters that also provide light; that is especially true when talking about halogens. What is overlooked in most discussions are the true life cycle costs for the different forms of lighting. In colder climates, where in the past incandescent lighting was providing incremental heating, a switch from incandescent to fluorescent can actually increase greenhouse gases, increase total life cycle costs, and therefore have yet another detrimental effect on the environment and the economy. Assuming one could continue to use incandescent lights and the facility was in the appropriate climate, such as those with four seasons, above certain altitudes over a vast portion of the globe or below certain latitudes, incandescent lighting might provide the best set of trade-offs. If that is true, a lot of research and a lot of engineering has to go into the actual facility to maximize benefits and minimize costs. With sufficient opportunity to help meet a heating load, even seasonally, incandescent lights may make a much better choice than business has been led to believe, and unfortunately what government and environmentalists think they understand. All sorts of facilities closer to the equator must look at LEDs as the preferred alternative, based on science and not the propaganda put out by many organizations, including governments. Switching from ILBs to CFLBs may not always result in an environmentally friendly outcome, especially in cold climates, as the authors of a Canadian report conclude: "In summary, switching from ILBs to CFLBs may not always result in an environmentally friendly outcome, especially in cold climates" (Ivanko, Waher, & Warbody, 2009).

Heating is another extremely complex area; for the vast majority of businesses it can be placed lower on the list. Realistically, individual operations are going to have to make a number of different choices about heat.

The primary choices are either electrical heating or some form of fossil fuel system. Based on the technologies available in 2012, choices for many smaller businesses in their own facilities are extremely broad. For those in leased or rented spaces, utilities may or may not be included in such payments, and therefore choices may be more limited. For those who are planning on using leased or rented spaces, consideration of heating and cooling sources should be a factor in selecting a location and might even need to be looked at from an economic versus environmental cost trade-off standpoint. For those who are actually going to design or build their own new facilities, both geothermal and solar systems offer reasonable alternatives either as primary systems or as auxiliary/backup systems. The bottom line is that from an environmental and life cycle standpoint, natural or liquefied petroleum gas offers significant advantages over all other forms of fossil fuels. Today, additional sources are being developed that might replace any or all of these depending on the size of the operation, but all need to be considered and the positives and negatives in terms of hazards to the air hazards to the Earth and waste streams have to be examined. The larger issue is what sort of operations go on in the business and whether there is a way to recover large quantities of otherwise wasted heat to offset the need to generate additional heat in order to maintain a viable internal environment. As an example, and a historic reference point, in the days before digital electronics, it was not unusual for military communications facilities to be designed to recapture the heat from tube-operated equipment to maintain reasonable operating temperatures within their facilities. The Naval Communication Station in Adak, Alaska had a very aggressive plan to recover heat from a lot of the tube-operated terminal equipment to help maintain the temperature in their facility and reduce heating costs.

Various forms of geothermal heating and cooling are also available, but there are many variables, which include water tables, quality of water, overall installation, and maintenance costs of such a system and the recognition that by strict definition heated water becomes a regulated waste stream because heated water pumped back into the environment can have significant impact, on groundwater. The impact may or may not be minimal for water pumped back into bodies of water, but the impact on the entire water bodies' ecosystem can be significant. We will come back and look at some additional opportunities for capturing "heat" associated with any number of operations. When we start talking about heat recovery and air distribution, something as simple as effective ventilation allows the recovery of a tremendous amount of heat value in production and related operations that in turn can help reduce the "costs" of heating.

Cooling offers fewer opportunities today simply because of the laws of physics and the current state of technology. However, geothermal cooling and simple wet cooling properly deployed in appropriate climates offer opportunities that are seldom considered, but with the rising cost of fossil fuels, increased regulation, and the economic cost of regulatory compliance, these are areas that are also going to need further study. One big issue with cooling when one gets beyond a certain facility size is that many current techniques use chemicals that are hazardous, and they become a challenge to manage, represent increased liabilities, and in most cases create hazardous waste streams: ammonia is one example.

For very large facilities or for larger properties such as schools or school campuses, large office parks, prisons, ports, and similar operations, it is not unusual for the occupying entity to build its own power generation facility and for larger properties, water treatment and sewage treatment facilities. Based on current technology, new forms of energy generation including such technologies as pebble bed atomic reactors might make sense. Many power generation processes depend on the heating of water, and that represents an additional set of risks liabilities and waste streams. Factors often overlooked in the development and management of heavy industries/large commercial facilities are the costs, dangers, and waste streams associated with boiler water treatment chemicals, scrubbers, and other peripherals and support operations required to maintain such a facility.

6.2.2.5 Utilities and Space

Except for fairly large facilities, utilities are most often purchased, so there is little in the way of hazardous materials or waste streams especially waste streams of airborne or waterborne emissions. That leaves the issue of space. This becomes another set of balance or trade-off equations: should one build or outsource production or should other inherently dangerous processes be isolated from or integrated into the overall space? How does building design and construction impact "livability" and adequate physical working space? Can effective use of any number of different techniques reduce the carbon footprint, the physical footprint, or the environmental footprint of a facility? One area that will receive more attention is the use of pervious rather than impervious materials for parking spaces. Another will be use of natural light to supplement electrical lighting, another is proper siting on a lot or proper selection of a lot to allow for proper siting and the need to plant or take advantage of existing tree lines based upon heating, cooling, and natural weather patterns. In essence, what we are saying is that any way you can reduce the need or use of generated power is a way to improve bottom line while at the same time providing for more environmentally friendly operation. Since we are taxing on the actual impact on the environment this is not an area that should be ignored although in many cases it may be an area that falls outside of the organization's ability to impact change.

6.2.2.6 Cleaning Supplies and Disposal Methods

Most people today recognize that the range of cleaning products and, to a lesser degree, such articles as oily rags represent another area of risk and environmental contamination. There are far too many products that carry far too many risks for us to cover in this rather high-level overview. The primary issue here is that a very large number of cleaning supplies represent hazards of one sort or another, and there is an even greater danger in mixing immiscible or chemically reactive cleaning materials. Many facilities that require rotating machinery and use petroleum based lubricants do not discard oily rags and other materials that absorb lubricants or solvents. There are a host of companies today that either manage or recycle such industrial effluvia. The important point is to recognize the inherent risk associated with the materials and then manage that risk. See some of our discussions to follow.

In summary, there are far more hazardous materials and possible hazardous waste streams in the typical office environment and in the janitorial arena than most businesses ever consider, and most do not give enough thought to this. When one starts looking at the incremental volumes and risks from all operations, from the simplest office operations to the most complex production operations to the most difficult process maintenance procedures, the total volume of materials that would need to be examined is huge. Chemicals and substances of all sorts are subject to regulation and therefore must be considered or treated "regulated" materials. All are subject to some form of control, and when all areas are recognized, both the number and the quantity of such regulated materials grow quite large and that does, in fact, represent a measurable and quantifiable risk to the organization.

6.2.3 Logistics Integration

Tables 6.2 and 6.3 help underscore the point that hazardous material exists all around us and that in general very few people have any awareness of, still less an understanding of, the hazards that surround us at work, at home, and in between. This is a critical issue for NGOs accepting postpurchase goods, where there are many "hidden" hazards that a consumer does not have to consider or deal with but that an NGO, as a shipper, must be aware of and manage. Many everyday items are hazardous material or can easily combine with other materials to create situations as hazardous materials. *Chlorine bleach* and *ammonia* combined produce highly deadly chlorine gas. Hydrogen peroxide as a common use antiseptic in a very dilute solution: 100% hydrogen peroxide (CLASS 5.1 OXIDIZER and CORROSIVE) must be chemically stabilized before it may be moved.

This table appears for a second time in this chapter because it is imperative that donor operations work hard to eliminate hidden or unrecognized regulated materials, while it is equally important to make sure that such items are properly identified and prepared for shipment when their inclusion is critical to the mission.

Table 6.3 General Use and Household Hazard Identification Matrix

Material	*Class*	*Example*
Calcium hypochlorite	Oxidizer	Algicide, swimming pools
Sodium hydroxide	Corrosive	Detergent and soap (TLV)
Sodium hypochlorite	Corrosive	Swimming pools, laundry
Nail polish, polish remover, perfume/cologne, most stains, and some paints	Flammable liquids	Self-explanatory
Liquid flavoring extracts	Flammable liquids	Vanilla extract, lemon extract
Herbicides, pesticides, fungicides	Toxic	All those gardening chemicals
Cooking, heating, or lighting devices fueled	Flammable	Coleman® lanterns, Coleman stoves, immersion heaters
Charcoal	Flammable solid	Self-explanatory
Untreated raw wood	Invasive species	Firewood, volunteer constructed wood pallets, unapproved blocking, bracing, and dunnage
Fire extinguishers and hair spray	Compressed gas	Self-explanatory
Correction fluid	Toxic	Office
Toner	Flammable	Office
Ink pen cart	Flammable	Office

Table 6.3 *(Continued)* General Use and Household Hazard Identification Matrix

Material	*Class*	*Example*
Printing inks (older formulations)	Flammable	Local sign and printing companies
Brakes	Asbestos	Cars and trucks
Smoke detectors	Radioactive	Home and office
Fluorescent bulbs and starters	Toxic	Everywhere (school)
Older fluorescent ballast (PCBs)	Miscellaneous (carcinogenic)	Self-explanatory
Batteries	Corrosive and explosive	Everywhere
Markers	Flammable	Office and class room
PC CRTs media hard drives	Toxic	Chromium, beryllium phosphors
Oxygen generators (all types)	Oxidizer	Aircraft and personal medical use
Airbags	Explosive and corrosive	Cars and trucks
Lithium-ion batteries	Flammable solid	Consumer electronics, toys, office devices, aircraft black boxes, and much more

6.2.3.1 Transportation Packaging

We will use slightly nontraditional terminology here in the hopes of maintaining clarity for all those who are not packaging professionals, and are not interested in becoming packaging experts. Technically, everything that contains, restrains, protects, displays, promotes, and adds to product usability or value at the user purchase level is "packaging." In a more focused framework, packaging is most often described as serving four functions and falling into three groups. The three groups generally used to categorize the entire packaging universe are: primary packaging, secondary packaging, and tertiary packaging. We are going to reduce that to two broad areas: primary packaging, that which occurs as part of the production process, and in many cases as an integral part of production (think of individual packets of salt, ketchup, or mustard, and of the individual dosage packaging of both over-the-counter and prescription drugs as examples where packaging is essentially an integral part of the product), and, transportation packaging, which is applied or conversely removed at the shipping and receiving areas of organizations.

Primary packaging may be an integral part of the production process or may occur at the very end of the production process so that it is an integral function of the product characteristic; and quite frequently initial packaging is literally contiguous to the end of the production process, so it will be addressed in the final segment (soda, beer, soup).

Logistics is involved with transportation and distribution, so transportation packaging will be addressed here. As part of the overall product development cycle and design process, there are issues that need to be addressed with logisticians and packaging specialists throughout the entire product development process. These decisions will directly impact on choices about specific technical functions for production line configuration, product package, product packaging, and transportation packaging.

> Transportation packaging has one overarching mission: preservation of the product from point of manufacture to the customer. This packaging adds to the value of the product in many direct and indirect ways. This includes the shipping container, interior protective packaging, and any unitizing materials for shipping. It does not include packaging for consumer products such as the primary packaging of food, beverages, pharmaceuticals, and cosmetics. In the United States, transportation packaging represents about one-third of the total purchases of packaging; the balance is attributable to primary packaging for consumer products. To design a transport package one must have goals or objectives in mind. These will vary with products, customers, distribution systems, manufacturing facilities, etc., but most transportation packaging should address the following:
>
> **Ease of Handling and Storage**—All parts of the distribution system should be able to economically move and store the packaged product.
>
> **Shipping Effectiveness**—Packaging and unitizing should enable the full utilization of carrier vehicles and must meet carrier rules and regulations.
>
> **Manufacturing Efficiency**—The packing and unitizing of goods should employ labor and facilities effectively.
>
> **Ease of Identification**—Package contents and routing should be easy to see, along with any special handling requirements.
>
> **Customer Needs**—The package must provide ease of opening, dispensing, and disposal, as well as meet any special handling or storage requirements the customer may have.
>
> **Environmental Responsibility**—In addition to meeting regulatory requirements, the design of packaging and unitizing should minimize solid waste by any of the following: reduction-return-reuse-recycle.
>
> Since transport packaging should always be economical, the above goals should be balanced or optimized to achieve the lowest overall cost (McKinlay, 1999).

In 1914, American railroads, which at the time were carrying most of the freight in the United States, recognized and authorized the use of corrugated and solid fiberboard shipping containers for packing many different types of products. Motor carriers, in turn, followed the railroads' example in 1935 when they adopted their own packaging rules that often called for fiberboard boxes. This standard packaging, employed during the early months of World War II, and specifically the colossal failures of packaging and the impact that had on the amphibious landings at Guadalcanal in 1942, which demonstrated how critical proper packaging design really is, forced the military to conduct their own research and create their own organizations to address packaging and eventually led to military specifications (MILSPEC) and military standard (MILSTD) packagings. The universally accepted packaging available at that time had proven adequate for rail and highway movement within the 48 continental United States but was grossly inadequate for the movement of goods overseas and into different and changing environments and prolonged exposure to the elements (Maloney, 2003). The scope of this disaster and the lessons learned led to the establishment of the School of Military Packaging Technology, the school disappeared from Aberdeen Proving Ground in October of 2009 as a result of BRAC, and some of this mission was transferred to the Defense Ammunition Center. Part of the reason this occurred was the rapid change in technology, specifically materials technology and packaging technology, which raised the overall standards of commercial packaging as well as military packaging.

The three biggest threats to product and packaging from a transportation and distribution standpoint in descending order are moisture, temperature, and pressure. If retail unit packages, which may contain multiple quantities of individual products or multiple packages of multiple quantities of products, include sufficient protection against moisture, there is less need for either more expensive for additional transportation packaging. On the other hand, as an example, if the organizational concept requires all outbound shipments be properly shrink-wrapped with the appropriate shrink wrap material, then theoretically the need for vapor barriers and desiccant materials may be reduced or eliminated. That might be particularly true for products that are produced for local or regional consumption and are yet still produced in relatively large quantities. This is why logisticians need a broad background than most recognize, and why logisticians need to be involved very early in the product development stages. Most texts on packaging identify four primary functions, although they may use slightly different terms or variance. The four functions of packaging are contain; protect; facilitate handling; and promote sales (Ramslund, 2000).

One area worth discussing briefly is packaging materials and packaging quantities. For those who entered the workforce in the 1960s and 1970s and have been involved with business-to-business commercial and industrial products, it comes as no surprise that packaging is a major issue. For those who study retail markets, consumer behavior, demographics, and psychographics, it is not news that packaging at the individual unit level for retail products presents serious challenges to getting the right balance between product protection and preservation versus safety versus ease of opening. The need to balance all three issues presents a myriad of challenges and has led to well-documented consumer frustration dissatisfaction while adding to product costs as well as waste stream growth.

The many issues related to movement of goods from point of production to point of sale is another area that represents general waste issues, damage and loss issues, and environmental issues and impact. While not really addressed here, the whole new field of reverse logistics applies the moment goods leave the point of production. The more important point here is that the concept of reverse logistics must be engineered into the product, the production line, and the packaging function. This is necessary to address the growing legal ramifications associated with the production processes that generate hazardous waste streams and emissions as well as the ultimate ownership and responsibility of goods that are, or contain, hazardous materials.

The types of material used for what is known as secondary and tertiary packaging, the methods used to secure or "build" the packaging, and, the methods processes and materials used to generate visual information on the packaging all can add hidden costs that represent hazardous material exposures and increased packaging disposal costs. Some examples of specifics are adhesives used in building the physical package, or methods of closure/sealing; inks paints or other materials applied to the packaging to provide visual appeal or critical logistics and transportation information; preservation of the package itself application of fumigants to tertiary packaging and containers; and passive and interactive RFID tags.

One often-overlooked area that the text will not address in detail but is worthy of separate discussion is invasive species and environmental issues related to the movement of goods by sea from one ecosystem/environment to another ecosystem/environment that is geographically removed by great distances.

Kudzu, Asian carp, Zebra mussels, and long horned beetles are just a few examples of invasive species that have drastically changed or are currently changing the local ecosystem. The most egregious example of such an unplanned consequence from modern air and sea transport is Guam; unfortunately it is one that very few Americans, including businesspeople, educators, legislators, and logisticians, are aware of.

It is now generally accepted that Americans unknowingly and inadvertently introduced brown snakes onto Guam in their post–World War II efforts to rebuild the island's economy. Those brown snakes have wiped out all native birds on the island, and it is now a full-time and not inexpensive job to manage and reduce the snake population on the island. There are ongoing efforts to restore an avian population to the island, but it will be an expensive and extensive long-term proposition.

While less pressing, those issues apply to commercial and noncommercial movement of goods and the transport devices (including those used strictly for pleasure) between areas within the United States. At the current time, because of an invasive species in parts of Worcester County, Massachusetts, raw forestry products, including firewood for personal use, cannot be moved out of the county for any purpose such as camp firewood per trip to another area. If you take the time to look at signs posted at many public boat ramps throughout the United States, you will notice that there is both a request, and a legal requirement, to properly clean or purge boats when they come out of the water even for movement within the local areas within the United States. An overlooked partner in this area is USDA APHIS. Agricultural products may not freely move into and out of every state because of the organisms they might carry, whether that is some form of plant disease such as blight or some sort of living insect such as arachnids or nematodes. Arachnids and nematodes love to travel in/on untreated wood, so we now have international phytosanitary standards. In the past, treatment of wood products was (or in some cases still) uses what we now recognize as extremely toxic/poisonous materials as an alternative to other forms of treatment. Recent cases demonstrate that under certain conditions the treatment of such packaging materials used in transportation, such as pallets, dunnage, and blocking and bracing can contaminate the product being transported.

As you can see, there is a lot more to product packaging, packaging for transportation, and transportation itself that needs to be considered and discussed. We will come back to that later, but now let us move on to packaging directly related to the product rather than its shipment.

6.2.3.2 Material Handling Equipment and Material Handling Systems

If you have material handling equipment, then you have hazardous materials issues as well as environmental health and safety issues. This is not a bad thing, but if you have not properly addressed all the requirements or you simply did not understand that there were so many requirements, this could be a weak area for many operations.

All such items use some form of motor, which is rotating machinery, so immediately there are personnel safety issues and lubricant issues. Internal combustion engines, regardless of the fossil fuel, create additional series of challenges based on the fuels themselves, air monitoring requirements, and circulation considerations. Electric-powered units have batteries. Depending upon what kind of batteries are being used, battery charging and servicing stations as well as other unique EHS requirements that might include acid absorbent spill kits and neutralization materials, eyewash stations, open flame and heat restrictions, and possibly special fire suppression and fire-fighting materials and protocols are needed. All represent additional costs, risks, and exposures. Both industrial and small lithium-ion batteries have been known to cause fires, so adequate research needs to be done both on the batteries themselves and charging systems that will be used, including the possibility for the need of fire suppression and current limiting devices. If you use lead acid batteries to start your material handling equipment, then you will need eyewash stations and servicing stations with adequate ventilation and personnel protective equipment. For an operation using rotating machinery, including typical MHE, lubricating products as well as lead acid batteries, are part of the package. There are two different sets of spill response/control items and two separate sets of requirements for safety materials and personnel protective appointment.

The larger issue is recognizing what hazardous materials are inherently parts of the material handling equipment. One category of hazardous materials must be kept on hand to service and support the material handling equipment and the kind of waste streams you are generating. There is a second set or group of materials that represent recognized and hidden waste streams. This includes everything from coolant to battery acids drained or unused excess oils and, of course, oily or otherwise contaminated rags. Solvents used to clean the equipment itself as well as the tools in the work areas and various preservatives may need to be treated as hazardous materials. Many service and maintenance aids that might not be treated as hazardous materials initially will have to be treated as hazardous wastes once they have been opened/used when disposing of them. That includes even the smallest quantities in the bottom of cans, drums, tubes, swabs, or brushes that have any foreign materials adhering to, or absorbed in, them. Some of the materials used in operating and maintaining material handling equipment are essentially incompatible, which means they must be segregated and that special efforts must be made to properly segregate those waste streams as well.

6.2.3.3 Transportation Methods

Most purchasers of transportation services and the vast the bulk of the transportation operators, since the majority are single-mode operators, usually look at the cost, time, and convenience factors when making decisions in regard to mode selection. Today, informed businesses do recognize all the impacts of life cycle management and are also looking at other costs and impacts in deciding on their total distribution chain even if that function is exclusively provided by 3PLs and outside contractors. Energy efficiency and CO_2 emissions are both factors that need be applied when designing the distribution system. A properly designed system can minimize costs to the company in relationship inventory cost; convenience for the customer, or rapid delivery; and green concepts or impacts. The chart below (Table 6.4) clearly demonstrates that there are a number of factors that need to be considered by both transportation sector providers and users when making decisions in regards to transportation and distribution models.

For many reading this, a lot of the information in the table above is superfluous to the point of overkill, but for certain industries and different specific businesses and sectors all of the information has value and impact.

Realistically, very few modes of transportation can provide for uninterrupted door-to-door movement of goods from place of production to place of sale or use (the two are not necessarily the same when dealing with retail products). That means almost all goods will complete some part of their journey by truck. That by no means suggests that truck transportation is the best, or even a good, choice. In addition, if one understands the transportation industry, even for a material that travels exclusively via truck, it is likely that the material will flow onto at least three and possibly as many as seven or more different transport units before arrival at its final selling point. Many producing organizations capture the major issues related to operating their own fleets or contracting out to 3PL or other external provider. What few recognize are the huge chemical costs related to the service and maintenance of a vehicle fleet. All information that applies to material handling equipment above applies to an organization that wishes to operate its own vehicles or decides to maintain such a fleet.

Table 6.4 Modal Comparison Environmental, Safety, and Health Factors

Mode	*Energy efficiency index*	*Speed*	*Average haul*	*Deaths per billion passengers (kms)*	*Date introduced*	*Vehicle life (years)*	*CO_2 emission (gms per ton per km)*
Air	1	400	1000	0.02	1958	22	540
Truck	15	55	265	2.4	1920	10	50
Rail	50	20 (200)	500	0.55	1830 (1970)	20	30
Barge	64	5.5	330	Very small	17th C	50	
Pipeline	75	4.5	300	Negligible	1856 (1970)	?	
Liner	100	16.5	1500	1.0	1870 (1970)	15	25

Source: Alderton, P., *Port Operations*, Informa, London, 2008.

It takes a careful examination of the information in the table to make decisions. And for many, the decisions are going to be driven by other factors. It is only when we start aggregating large quantities and talk about initial distribution that a lot of this information comes into play. Nonetheless, it is critical that the issues represented by the information in the table be considered when making large-scale, and to a lesser degree, smaller-scale or local, transportation mode decisions.

Many travel booking services now provide CO_2 emission figures as part of the reservation package for air travel; this is shown as how many tons of CO_2 emissions each flight leg of a trip represents. Given the direction of global regulation and taxation, the concept represented by that information now carries an increasing direct economic burden, whether that is viewed as a hidden cost, and, whether or not this is passed on to the ultimate consumer of the good or service. There are inherent safety and cost considerations related to each mode; based on either, it is fairly easy to create ordered lists to help choose transportation modes for the movement of goods. Reliability is another parameter; however, two parameters are becoming more important: energy efficiency and emission generation. With careful planning and attention to detail, the supply chain manager can build in improved energy efficiency and reduced emissions to reduce the cost to the producer and consumer. At the same time, wise choices and careful planning within the supply chain also contribute to the reduction of environmental impact. In the end, based on current trends, society as a whole is going to have to understand and then be willing to “pay for” the environmental impact of the various modes of transportation, as well as the environmental impact of the various modes of production and packaging. Businesses are going to have to accept and understand the negative impacts of industrial processes, which include production and transportation, upon the environment and then decide how to either avoid or allocate those costs.

While there are no absolutes, some basic decision points can be presented here. For those concerned strictly with local operations, the chart above has little practical application. For those who are shipping strictly within a region, the chart begins to take on the meaning but the options and choices are still limited. For those distributing nationally or globally, the chart begins to take on a lot more importance. Realistically, both those who are shipping and shipping companies themselves need to pay particular attention to the first three columns and the last column in the chart.

In addition to that, companies distributing beyond the region must factor in the issues of warehousing and distribution centers. Today, the majority of business texts treat the transportation distribution model as a trade-off between transportation costs and distribution center/warehouse costs and convenience. The big issue is the cost of inventory sitting in warehouses and distribution centers as well as cost of inventory in transit. That is an excellent model and is still the foundation for such discussions. Today, however, driven by the growing body of scientific research and the increasing tendency of governments to regulate and impose "carbon taxes" domestically as well as globally, it behooves both the business community at large and the transportation community at large to look more carefully at some of the information provided in Table 6.4.

Today, the final decision must be driven by a combination of company culture, interest in, and legislation encouraging "green" approaches to business; consumer perceptions; and, cost benefit analysis that trades off time versus delivery cost versus convenience. Make no mistake, for any entity that is delivering large quantities outside of the region, all those issues must be taken into consideration.

6.2.3.4 Transportation Providers and Protocols

If one looks at inbound logistics for goods-producing organizations, time cost, liability, and reliability all become drivers. Often, adjusting overall production schedules may allow for a move to less expensive and less environmentally unfriendly modes of transportation. That is a key concept in this rather short section. The just-in-time approach to manufacturing or production does not necessarily or automatically negate the use of less expensive modes of transportation. Typically, less expensive modes of transportation also represent those modes with the lowest carbon footprint and environmental impact. The trade-offs include time as an absolute, reliability, and seasonal impact on the mode of transportation. Something as simple as scheduling pickups or deliveries at the manufacturing facility for "off" hours can have significant direct and indirect environmental and economic impacts. Time of day pricing for surface transportation is now embedded in the Southern California port regions and can be expected to expand throughout the United States in all major port areas.

Commercial trucking operations and rail operations can have significant impact on congestion as well as "pollution." If we parse the normal 24-hour day into three "shifts," one could make the case for shipping and receiving operations being restricted or focused upon what is traditionally labeled as the "midwatch" by the military. That could be an arbitrary 8-hour band of time, for example, from 8 p.m. to 4 a.m. Taking the theoretical step further, shipping and receiving areas can be physically isolated from the rest of the facility; then there might be additional heating and cooling cost reductions associated with operating such facilities when the rest of the facility is closed down. This is not to suggest that any work has been done to quantify the advantages and disadvantages of such a concept. Humans are "wired," if you will, to operate during the day and sleep during the night, so safety in terms of driver ability/capability is one of the many factors that would have to be looked at under such a scenario. From a pure transportation cost basis and a pure environmental impact basis, there are a tremendous number of positives to exploring such an approach.

In addition to mode selection and time of day decisions, another major decision for many organizations, one that should help drive the outsourced transportation function decision, is selecting an appropriately equipped and operating carrier. Here the issue is less a hazardous material or waste material consideration than it is an environmental impact safety and reliability issue.

The last piece of the transportation issue is packaging. While not a major consideration, the total distribution chain for product will drive packaging design and packaging decisions. Proper choice of primary and secondary packaging, coupled with the intrinsic characteristics of the product, may allow for consumer products to move with equal reliability and safety via any mode of transportation. On the other hand, larger industrial products that might not ordinarily be considered "fragile" might need additional packaging for ocean voyages, especially since today a very large number of such items are "modularized" so they may fit into shipping containers.

6.2.4 Client Services

6.2.4.1 Understanding of Regulatory Environment and Constricts in the Field, and Nation State Differences

This is probably the big one in terms of hazardous material life cycle management and client services. Many NGOs, both large and small, respond to international disasters, often with very little advance notice or preplanning. It is critical that they have the expertise or access to expertise that assures them that they will be operating legally and in a culturally acceptable manner in all areas related to the use and disposal of hazardous materials. Generally acceptable practices in their home location may be adequate for deployment on the same continent, but serious issues can arise if it becomes necessary to move people and material into the international space. One specific concern is a failure to meet the stricken nation state's requirements. There is a triple or quadruple effect here, and this is the author's concern. Assuming that dangerous goods have been shipped and have arrived at their destination country, if there are serious enough issues with that material, it will never reach its final destination; that is problem number one. The sponsoring organization is now short of critical materials at the site where they are most sorely needed, which in turn means that materials will have to be procured. If those materials are procured globally or through local sources, the reality is that they are likely to cost more and will not necessarily be of the same quality. The next issue is fines that might be imposed, which in turn are coupled to the legal costs to defend and the negative impact on the sponsoring organization's image and additional friction between the sponsoring organization and the host country. That still leaves the dangerous goods "frustrated" at a foreign port, so the sponsoring organization faces the choice, assuming it is allowed to, of exporting back to country of origin, or disposing of the materials as regulated wastes if it is allowed to do that, or it may have the opportunity to sell the goods as distressed goods to someone else for him or her to do with as he or she sees fit. Beyond those issues are the issues of shelf life. Being right and winning on the legal/political playing field is still not enough if organizational time and dollars have been invested in fighting to win only to discover the material is no longer usable. Again, we end up with waste materials which, if you remember from earlier reading, are essentially still owned by the sponsoring agency. In some cases, the issue may be as simple as getting the correct adapter fittings to allow American-built units to interface with locally built hardware.

Actual environmental and health exposure and liabilities are relatively small here, but legal and cultural barriers can create huge problems in terms of procuring critical materials in a timely manner, time and effort diverted from the mission into the resolution of conflicts around these dangerous goods, and, issues relating to international relationships, image, and reputation.

6.2.4.2 Training and Credentialing of Field Staff and Volunteers

Many of the issues touched on above apply to this area as well. In the United States, we do not have licensure for those who certify dangerous goods for transportation. In some countries, one may run into licensure requirements for those who prepare and ship dangerous goods. In addition, there may be licensure requirement for those organizations and individuals offering the courses needed to earn licensure. By the same token, there are the equivalent of reciprocity agreements through the two major international regulatory bodies. An IATA-approved course from an approved instructor in one country is more often than not going to be recognized in another country. The same holds true for IMO-approved instructors and therefore the reciprocity for the training received.

There needs to be considerable effort made to ensure an agency knows whether or not it will have to handle any dangerous goods, and if so, the agency needs to ensure it knows all the requirements in regard to dangerous goods, so it can either obtain the necessary training or withdraw from areas where its staff and volunteers do not meet the host countries' requirements.

6.2.4.3 Understanding of Practices and Materials to Be Avoided in the Field Environment

Here, the issue revolves more around environmental, health, and safety issues; a full understanding of geographic and geologic issues in the area where one is operating; and an understanding of cultural issues. Gasoline and diesel are used in many developing countries, but supplies are so limited that theft is not unusual. In such a case, assuming there is an option, then one would want to look at the possibility of some form of gas as fuel for cooking or even boiling water. A country's or region's ability to process and treat hazardous waste streams also must be taken into consideration: what might work in a large city almost anywhere in the world might be quite inappropriate for a rural setting. Packaging, including the innermost shell casing or package for the hazardous materials, must be considered. Here again, the issue of improper disposal can lead to contamination, overloading of the local facilities, considerable extra transportation costs for hauling the wastes away, or simply extreme unsightliness and degradation of the environment. An oversimplified example would be the use of citronella candles versus the use of other insect repellents that come in pressurized containers. A citronella candle burns down and leaves little if any waste of any sort and what is left is not a hazardous waste. Spray cans of insect repellent introduce chemicals into the local environment, and create a disposal issue for the "empty" cans and partially used cans of spray that are thrown out or will be left behind. Again, this sort of waste is a hazardous waste stream even if it is not regulated in the host country. Such personal wastes place an increased load on the host country's ability to process or recycle materials that are by definition damaging to the local environment.

6.3 Defining Organizational Roles and Responsibilities

For an NGO that will be delivering goods and services that include or use hazardous materials outside of the NGO facility itself, there are significant risks based primarily on the fact that such services are not delivered routinely or continuously. For routine operations, paid staff and regular volunteers can certainly handle many of the issues outside of the hazardous materials issues. If an NGO needs to ramp up on an entirely irregular and unpredictable basis and get involved with moving large quantities of people and goods, then it becomes necessary to make sure not only that roles and responsibilities are well defined but to make sure that training records, certifications, and credentials are maintained. Given the fact that dangerous goods and hazardous materials regulations change annually or biannually and that all those regulations specifically require training of "employees," the administrative burden grows quite large. The risks also increase unless the entire hazardous material management program is in place, reviewed very regularly, updated as frequently as necessary, and is exercised. The risks to the organization and the potential risks to the environment and to the public at large are so great that special effort needs to be placed on the environmental, health, and safety issues and requisite training and qualifications of individuals. While it may be quite reasonable to place the majority of the operational movement and deployment of hazardous materials within the volunteer structure, it is critical that leadership in top management of the NGO receive and maintain the appropriate levels of awareness training and that they exercise due diligence at all times in anticipation of unexpected demands. For some organizations, it may be possible to recruit both the management and the operational talent to address all hazardous materials issues for routine as well as emergent and quick reaction needs. In those cases, those at the very top must fully understand and take responsibilities in regards to hazardous materials, or the organization could be placing itself and its prospective clients at greater risk for direct and personal injury as well as significant legal actions in regards to hazardous materials and hazardous wastes. The language in 49 CFR part 172, subparts G and H, places very specific and extremely broad responsibilities on the employing organization in terms of communicating information and training individuals. Failures carry legal, financial, social, and environmental impacts. Because of the inherent nature of emergency and disaster operations, there is a much higher than normal need to manage, at the strategic level as distinct from operational/tactical level, all hazardous material. That focus is primarily in the area of training communication and awareness. It is logical to expect that everything related to hazardous materials and the associated environmental, health, and safety issues is identified as the responsibility of a senior full-time staff member and that the entire top management team can prove that they routinely receive appropriate training commensurate with the level and nature of their responsibility. Whether the hazardous materials management responsibility is assigned as a primary function for a senior individual or it is treated as a collateral duty for senior individual, there must be a clearly identified hazardous materials chain of management supported by significant administrative, training, and record-keeping resources.

6.4 Stakeholder Identification

It would be easy to say that everyone is a stakeholder when it comes to NGOs, and there is a certain amount of truth to that idea. However, if we are talking very specifically about life cycle management of hazardous materials, then there are a number of broad categories of stakeholders that the NGO must recognize.

The key stakeholders include the entire paid organization and all its volunteers. That can be a very large amorphous and changing population; it clearly includes those who provide the input resources; external state and international government entities and bodies; all commercial and often some military transportation providers; and the entire organizational workforce, paid or volunteer, as one autonomous group. That last group must be broken down into subcategories, including the frontline employees who must interface with the public on a daily basis as part of their job at the normal business places of the organization; those employees who have clearly identified impacts, and lasting influence on, the overall life cycle management of hazardous materials and the senior management team within the entity; those people who will be deployed to the field and will interface with the public and handle or influence the handling of hazardous materials or hazardous wastes; and volunteers. Outside of the entity itself, the major stakeholders are prospective donors or funding sources for transportation providers and governments. That is a very large and poorly defined set of stakeholders, which is one of the major challenges for the NGO.

References

Alderton, P. (2008). *Port operations.* London: Informa.

Bierma, T., & Waterstraat, F. (2000). *Chemical management.* New York: John Wiley & Sons.

Eartheasy. (2012). *LED light bulbs: Comparison charts.* Retrieved from http://eartheasy.com/live_led_bulbs_comparison.html.

Falkman, M. A. (2001, August 1). Packaging function, structure defined. *Packaging Digest.*

Griffin, R. D. (2009). *Principles of hazardous materials management.* Boca Raton, FL: Taylor & Francis.

IATA. (2012). *Dangerous goods regulations.* Montreal, CA: IATA.

ICAO. (2010).*Technical instructions for the safe transportation of dangerous goods by air.* Montreal: ICAO.

IMO. (2010). *International maritime dangerous goods code.* London: IMO.

Ivanco, M., Waher, K., & Wardrop, B.W. (2009). *Impact of conversion to compact fluorescent lighting, and other energy efficient devices, on greenhouse gas emissions.* Retrieved from http://www.intechopen.com/source/pdfs/11764/InTech-impact_of_conversion_to_compact_fluorescent_lighting_and_other_energy_efficient_devices_on_greenhouse_gas_emissions.pdf.

Maloney, J. C. (2003). *The history and significance of military packaging.* Fort Belvoir, VA: Department of the Army.

McKinlay, A. (1995). *Need a new direction for your transport packaging? Do a 180!* Chicago, IL: IOPP.

McKinlay, A. (1998). *Transport packaging.* Chicago: CRC Press.

Moll, E. (2012.). *Energy-efficient bulbs: Halogen vs. fluorescent vs. incandescent.* Retrieved from http://greenliving.nationalgeographic.com/energy-efficient-bulbs-halogen-vs-fluorescent-vs-incandescent-3228.html.

Puleo, S. (2004). *Dark tide: The great Boston Molasses Flood of 1919.* Boston, MA: Beacon Press.

Ramsland, T. (2000). *Packaging design: A practitioner's manual.* Geneva: International Trade Centre.

Threat Analysis Group. (2010). *Threat, vulnerability, risk—Commonly mixed up terms.* Retrieved from http://www.threatanalysis.com/blog/?p=43.

Chapter 7

Putting It All Together

7.1 Introduction

7.1.1 Understanding the Concept of Hazardous Materials Life Cycle Management

There is nothing new or unique about hazardous materials management or life cycle management. Unfortunately, the bulk, but not all, of the work on related topics has been separated into those two areas, and a tremendous amount of what is written about hazardous material management really focuses on the operational environment rather than on the product and the process. There are a limited number of programs and researchers tackling the issue, so most of the literature, and therefore the most popular best practices, are built upon a framework that assumes incidents will occur and we must focus on incident response, postincident cleanup, and minimization of incidents. Every bit of that is critically important, but it ignores the larger issue of how to reduce the overall use of hazardous materials, how to maximize the effectiveness of the materials used, and how to minimize or better utilize hazardous material waste streams. The total hazardous material impact from any business equals the total of a number of small incremental hazardous material uses and even smaller hazardous material waste streams. Therefore, it becomes necessary to examine the entire business unit, and it becomes necessary to more closely examine the product and processes used to make the product. Since hazardous materials exist in the office environment and in the janitorial/maintenance areas, this applies to services that are the produced as well as tangible products

A hazardous material life cycle management approach recognizes that every process and every material used in the production of finished goods represent an opportunity to reduce the usage of hazardous materials, move to less hazardous materials, or move to alternate processes. Therefore, life cycle management of hazardous material must include direct and indirect processes, products, and waste streams, as well as all support functions. In the everyday world, we tend to focus on regulations, compliance, costs, and "exposure" when we talk about hazardous materials. What is lost is the underlying reason that governments are forced to regulate: the inherent dangers to the global environment of excessive use and improper disposal of that which is hazardous. Going back to that very simple concept, it is easier to understand why life cycle management of hazardous material needs to focus on removing as much material as possible from the entire process, converting it to nonhazardous or less hazardous materials and processes, and, finding ways to recover hazardous waste streams and turn them into productive assets in one manner or another.

7.1.2 Understanding What Regulated Materials Are and Under What Sets of Conditions They Are Regulated

The United Nations Economic and Social Council Orange Book revision in 1996 and the recent issuance of the GHS is the first and most significant step toward aligning all the disciplines, all the different agencies, and all the different governments into a unified and universal understanding of, as well as definition of, hazardous materials, regardless of the field one is working in. The actual number of different materials, substances, and streams that are regulated is much larger than most people recognize. That leads to a tremendous amount of confusion, as well as promulgation of misinformation and disinformation by those who do not have a solid grounding in hazardous materials, or dangerous goods if you will, based on the bedrock material covered in the Orange Book.

Despite what many may think, regulations governing hazardous materials have been put in place to meet two key objectives: to protect the environment and mankind from ourselves, and to ensure the free flow of commerce. Many might take exception to the statement, but in the end that is truly the purpose of most regulations, and it is especially true regarding all facets of environmental, health, and safety issues. Whether nation states and international bodies have abused their power in enforcing those regulations is a whole different story, but it should be clear that there is a need to promulgate the equivalent of best or safest practices and provide guidelines to ensure the safe conduct of business. It is also important to recognize that, for the global community as a whole, there is a trusted source to learn what the guidelines are for conducting business in relationship to environmental, health, and safety issues.

That got us off the track a little bit, so let us come back to the concept of "regulated." The impact, or for that matter the visibility, of the regulations may be nonexistent when one gets to the individual consumer level for many products that show up in retail outlets such as grocery stores, drugstores, departments stores, and hardware stores or do-it-yourself centers. The government in the broadest sense recognizes that there is an inherent danger associated with those products, and that both production and shipping of quantities of those materials must be subject to all parts of the regulations. At the same time, retail outlets and retail buyers may be unaware that such materials are regulated and subject to specific requirements, even when/if they are thrown away or given away. This dichotomy will continue to exist for an extremely long time but that 1996 decision by the UN will eventually trickle down and allow professionals in what traditionally have been completely separate silos to start teaching and preaching from a common set of definitions. That in turn will make it possible to begin to address how to raise awareness within the general public so that people recognize, without being scared of, the hazards that surround them everywhere and all the time. For the foreseeable future, the current system still stymies those who study hazardous materials life cycle management in order to educate others and ensure that everyone does everything actually required by law to reduce the use of hazardous materials and thereby reduce vulnerabilities and waste streams.

7.1.3 Understanding the Concept of Hazardous Waste Ownership, Even if by a Different Name and Regardless of Nation State

This is a core competency and precept for hazardous materials life cycle management. Industrial processes generate all sorts of streams, not just a product stream. The reality is that today we understand that the large majority of those streams are hazardous by definition, and even those that are not hazardous are creating conditions that are hazardous to the environment. The simple answer is "recycling" and another term is *repurposing*; these are key components in true life cycle management of hazardous materials because they recognize the need to recover value from all waste streams, as well as working hard to minimize all waste streams. Please remember this includes all the products that consumers throw away after they have used them. The first half of life cycle management of hazardous materials is reducing their usage; in reality, there will always be hazardous materials used, and there will always be wastes. Because of that, the second half of this very complex equation is deriving additional value by effective use or reuse of all streams from all business activities: all the administrative and support functions, the production functions, and the new one to many reading this, the products themselves in the postpurchase environment all the way through to consumer disposal. As pointed out in the text, the current trend in global law and regulation is to hold producers of goods responsible for the life cycle of the product. Whether you want to talk about carbon taxes or deposits on bottles and cans, the most recent moves are toward assigning specific costs to producers that represent the municipal or societal costs to remediate product into the environment. The trend is clear. Ignoring the entire reverse logistics process and recycling sectors within the world economy, the key issue is that as a producer, and in a growing number of instances as a consumer, once you buy materials you own them for life whether hazardous or not hazardous. For hazardous waste streams, this becomes an accounting issue and a legal liability issue. Effective management of the total life cycle, so that products and component parts of products are returned to use in one form or another, becomes one of the drivers for effective life cycle management of hazardous materials.

7.1.4 Converting Hazardous Waste Streams Into Nonhazardous Waste Streams or Recycled Applications

This is really a continuation of the idea introduced above, an area where reverse engineering, industrial engineering, and best environmental health and safety practices meet. Metal cases can be ground and remelted to reduce the cost of new metal products. Many plastics and plastic cases can be ground and turned into fuel streams or raw stock for new products such as paving materials and outdoor plastic-based building materials for yards and gardens. Even old paving materials are now being recycled to enhance and extend new paving materials. In some cases, usable product can be reconditioned and introduced into new markets or secondary markets. The list is endless and the choices are varied, but the bottom line is that first one needs to understand what is hazardous, in all definitions of that word, in the finished product, and how that material can be made into a usable, and preferably nonhazardous, stream. Examples mentioned and worth repeating are using some hazardous materials or "contaminated" hazardous materials as feedstocks to produce heat or energy. If high enough temperatures are used, then in some cases the chemical process involved in that high heat essentially "neutralizes" the hazard. In other cases, if the hazard is primarily nothing more than the flammability of the material, then finding ways to enhance its value and recover the BTUs as a fuel is yet another opportunity. The bottom line is to reduce the number of streams, reduce the number of hazardous streams, and find ways to recover value and usage while decreasing risk by the effective redeployment of postproduction materials or production process materials.

7.2 Eliminating Silos and Stovepipes

This is not a new idea but another critical element in achieving the goals of hazardous materials life cycle management. In a large organization such as an automobile manufacturer, all of the following disciplines or professional specialists have roles to play: design engineers, test technicians, QA technicians, industrial security specialists, industrial hygienists, safety specialist, manufacturing engineering, plant maintenance engineering, safety specialists, machinists, materials handling equipment operators, shippers, receivers, industrial engineers, and finance and purchasing. External partners include all of the first response agencies that might be expected to provide services in case of a catastrophic hazardous materials incident, local government including code enforcement, planning boards and zoning boards, all suppliers of goods and materials, transportation service providers, capital equipment and rolling stock providers, and state and federal regulatory agencies. Effective management of hazardous wastes involves everyone, and there are costs that need to be captured and assigned, risks that need to be weighed, best business practices that need to be applied, and awareness of changes in technology and regulations (which in some cases may make a process or process chemical unacceptable) and many others.

The key here is that decisions cannot be reached in a vacuum—that they are neither purely economic nor purely safety nor purely environmental nor purely within the realm of the producing organization itself. The lack of a common starting point for the environmental, health, safety, and transportation disciplines in relation to hazardous materials issues has fostered the growth of separate and often conflicting silos. What might be exempt or narrowly defined under one set of regulations may be in conflict with another set of regulations. That sort of intrinsic conflict has led to many of the silos mentioned above, but the steps taken in 1996 by the UN to create, first, a knowledge base and/or definitions if you wish, to allow all these different areas to start coming together and to start working together. One of the biggest challenges the private sector and the NGO sector face is operating at the interfaces between various federal agencies, and the federal regulations. If OSHA, EPA, and DOT each have a different definition of "flammable liquid" or if only the United States recognizes the concept of "combustible liquid," then it becomes extremely difficult for multinational companies to achieve common goals across the global supply chain where products were bought in one country and shipped to another country, and where finished goods are shipped to many countries. The other challenge that businesses in the last part of the 20th century and the beginning of the 21st century have recognized, and are addressing, is that adversarial relationship between business partners does not benefit any party in the end. Both goods and service providers as sources of inputs to the enterprise as well as the entire distribution chain and then users on the output side need to be recognized as partners so that all those who make up the life cycle of a product become partners in the management of the product from its earliest stages to its final disposition.

7.3 Effective Planning and Preparedness

It is reasonable to assume that many hazardous materials and many processes using hazardous materials and generating hazardous waste streams will continue to be used through most of the 21st century. That is the reality and as such it must drive the planning effort and preparedness posture. Here we are speaking only about the reactive component or what can be called *incident management*. Incidents will occur for any number of reasons, and effective planning that leads to improvement and preparedness focuses on minimizing the impact or effect of those incidents. Given the shifting weather patterns and changes on the Earth, it is also reasonable to assume that events beyond the control of humans are going to conspire to create situations that will be serious in nature and might affect a single facility, a community, a region, or an even larger area. Again, it becomes the responsibility of the security or risk manager, the emergency managers, and the community, as well as, any in the organization, and corporate management team to ensure that plans are made to address these issues. These plans extend beyond the issue of hazardous materials into the area of continuity of operations or business continuity. Conceptually, these all must be treated together because the two highest likelihoods of business interruption are catastrophic events as a result of a hazardous materials incident, and a weather, or a geologic induced, event. Anything from earthquakes and tsunamis to sunspots to extended flooding, or, ice storms can render an individual geographic business unit inoperable while creating huge environmental, health, and safety risks. Emergency planning, by any other name, is an important part of the management function today. Proper planning coupled with proper training of the entire workforce can mitigate the impact of the inevitable, no matter how unlikely, catastrophic incident especially in the facility that routinely uses large quantities of hazardous materials. In the following text, we examine briefly two key components related to acute situations.

7.4 Developing Response and Emergency Action Plans

At the strategic level, planning encompasses much more than response plans and emergency action plans. At the operational level, these become two key tools to help minimize hazardous materials incidents and prevent them from moving from small localized emergencies into major disasters or catastrophes. The strategic plan would address corporate issues such as business continuity and the need for response and emergency action plans. At the operational organizational level, teams or groups need to be assigned the responsibility to develop such plans. Every fire drill represents an example of an emergency action plan. Every law enforcement EMS or fire response to a facility represents what should have been an exercise of emergency action plan. An emergency can be a single person having a heart attack, and while most would not think that calling 911 is an emergency action plan, that action does represent an emergency action plan. As industrial and business processes become more complex, the need for emergency action plans and the breadth of the facility that needs to be covered by the emergency action plan expands exponentially. For many organizations and in many ways, response plans and emergency action plans are very similar and one might even look at the response plan as being the very function of external governmental bodies. An emergency action plan, on the other hand, is the response within the organization. That is one way to distinguish the two, but there are elements of both types of planning on both sides. That becomes more apparent in larger organizations which have single facilities covering large areas of land where they may maintain their own security fire and medical emergency response teams. Depending on the nature of the organization, and the nature of the materials on hand, it may also make sense for an organization to train an internal response group. Typically, such a response group is a first line of defense in a tiered active defense. That group has initial responsibility for trying to contain the incident and to evacuate people. Typically, it takes assets beyond the employees within the organization to properly or fully contain a hazardous materials incident, and in the vast majority of cases it will take contracted outside specialists to fully mitigate and restore the site. All that might require moving a company's production off-site and all of that might also require already executed contracts for repair and restoration services. That represents all the steps that must be followed for full mitigation.

7.5 Exercising Plans/Tabletops/Training

There are two golden rules in emergency management, and they apply here as well. Rule one goes something like this: "it is not planning that is critical, it is the planning process." Rule number two is just as straightforward: "After an incident has occurred is the wrong time to get to know those individuals who are going to come to your aid." Knowing all of the responders and having agreements with them, including those in the public sector, is critically important and should occur as a result of the planning and response efforts noted above. Without the necessary training and the creation of scenarios where they are called upon to exercise training evolutions whether tabletops, simulations or full-scale exercises, success in responding to an incident only comes from practice, and practice requires commitment from management and an active program to develop exercises. It is also important to note that there may be situations, including emergent needs outside of your area or region, where the organization is tasked to support someone else. Partnerships are two-way streets and participating in other people's exercises as well as creating your own exercises is also an important part of developing those relationships locally and globally that will allow you to survive even the worst sort of incident within your own facility or within your own community.

References

Alderton, P. (2008). *Port operations*. London: Informa.

Bierma, T., & Waterstraat, F. (2000). *Chemical management*. New York: John Wiley & Sons.

Eartheasy. (2012). *LED light bulbs: Comparison charts*. Retrieved from http://eartheasy.com/live_led_bulbs_comparison.html.

Falkman, M. A. (2001, August 1). Packaging function, Structure defined. *Packaging Digest*.

Griffin, R. D. (2009). *Principles of hazardous materials management*. Boca Raton, FL: Taylor & Francis.

IATA. (2012). *Dangerous goods regulations*. Montreal, CA: IATA.

ICAO. (2010). *Technical Instructions for the safe transportation of dangerous goods by air*. Montreal: ICAO.

IMO. (2010). *International maritime dangerous goods code*. London: IMO.

Ivanco, M., Waher, K., & Wardrop, B.W. (2009). *Impact of conversion to compact fluorescent lighting, and other energy efficient devices, on greenhouse gas emissions*. Retrieved from http://www.intechopen.com/source/pdfs/11764/InTech-Impact_of_conversion_to_compact_fluorescent_lighting_and_other_energy_efficient_devices_on_greenhouse_gas_emissions.pdf.

Maloney, J. C. (2003). *The history and significance of military packaging*. Fort Belvoir, VA: Department of the Army.

McKinlay, A. (1995). *Need a new direction for your transport packaging? Do a 180!* Chicago, IL: IOPP.

McKinlay, A. (1998). *Transport packaging*. Chicago: CRC Press.

Moll, E. (2012). *Energy-efficient bulbs: Halogen vs. fluorescent vs. incandescent*. Retrieved from http://greenliving.nationalgeographic.com/energy-efficient-bulbs-halogen-vs-fluorescent-vs-incandescent-3228.html.

Puleo, S. (2004). *Dark tide: The great Boston Molasses Flood of 1919*. Boston, MA: Beacon Press.

Ramlsand, T. (2000). *Packaging design: A practitioner's manual*. Geneva: International Trade Centre.

Threat Analysis Group. (2010). *Threat, vulnerability, risk—Commonly mixed up terms*. Retrieved from http://www.threatanalysis.com/blog/?p=43.

UN. (2009). Recommendations on the transport of dangerous goods. New York: UN.

USEPA. (2012 February 6). *Brownfields and land revitalization*. Retrieved from http://epa.gov/brownfields/.

Index

T

U

V

W

Z